Environmental Science II Lab Manual

Custom Edition for Lone Star College

Common Conversion Factors

Length

1 yard = 3 ft, 1 fathom = 6 ft

	in	ft	mi	cm	m	km
1 inch (in) =	1	0.083	1.58×10^{-5}	2.54	0.0254	2.54×10^{-5}
1 foot (ft) =	12	1	1.89×10^{-4}	30.48	0.3048	
1 mile (mi) =	63,360	5,280	1	160,934	1,609	1.609
1 centimeter (cm) =	0.394	0.0328	6.2×10^{-6}	1	0.01	1.0×10^{-5}
1 meter (m) =	39.37	3.281	6.2×10^{-4}	100	1	0.001
1 kilometer (km) =	39,370	3,281	0.6214	100,000	1,000	1

Area

1 square mi = 640 acres, 1 acre = 43,650 ft^2 = 4046.86 m^2 = 0.4047 ha
1 ha = 10,000 m^2 = 2.471 acres

	in^2	ft^2	mi^2	cm^2	m^2	km^2
1 in^2 =	1		—	6.4516	—	—
1 ft^2 =	144	1	—	929	0.0929	—
1 mi^2 =	—	27,878,400	1	—	—	2.590
1 cm^2 =	0.155	—	—	1	—	—
1 m^2 =	1,550	10.764	—	10,000	1	—
1 km^2 =	—	—	0.3861	—	1,000,000	1

Other Conversion Factors

1 ft^3/sec = .0283 m^3/sec = 7.48 gal/sec = 28.32 liters/sec
1 acre-foot = 43,560 ft^3 = 1,233 m^3 = 325,829 gal
1 m^3/sec = 35.32 ft^3/sec
1 ft^3/sec for one day = 1.98 acre-feet
1 m/sec = 3.6 km/hr = 2.24 mi/hr
1 ft/sec = 0.682 mi/hr = 1.097 km/hr
1 billion gallons per day (bgd) = 3.785 million m^3 per day
1 atmosphere = 1.013×10^5 N/m^2 = approximately 1 bar
1 bar = approx. 10^5 N/m^2 = 10^5 pascal (Pa)

Commonly Used Multiples of 10

Prefix (Symbol)	Amount	Prefix (Symbol)	Amount
exa (E)	10^{18} (million trillion)	centi (c)	10^{-2} (one-hundredth)
peta (P)	10^{15} (thousand trillion)	milli (m)	10^{-3} (one-thousandth)
tera (T)	10^{12} (trillion)	micro (μ)	10^{-6} (one-millionth)
giga (G)	10^{9} (billion)	nano (n)	10^{-9} (one-billionth)
mega (M)	10^{6} (million)	pico (p)	10^{-12} (one-trillionth)
kilo (k)	10^{3} (thousand)		

Volume

	in^3	ft^3	yd^3	m^3	qt	liter	barrel	gal. (U.S.)
1 in^3 =	1	—	—	—	—	0.02	—	—
1 ft^3 =	1,728	1	—	.0283	—	28.3	—	7.480
1 yd^3 =	—	27	1	0.76	—	—	—	—
1 m^3 =	61,020	35.315	1.307	1	—	1,000	—	—
1 quart (qt) =	—	—	—	—	1	0.95	—	0.25
1 liter (l) =	61.02	—	—	—	1.06	1	—	0.2642
1 barrel (oil) =	—	—	—	—	168	159.6	1	42
1 gallon (U.S.) =	231	0.13	—	—	4	3.785	0.02	1

Energy and Power

1 kilowatt-hour = 3,413 Btus = 860,421 calories
1 Btu = 0.000293 kilowatt-hour = 252 calories = 1,055 joule
1 watt = 3.413 Btu/hr = 14.34 calories/min
1 calorie = the amount of heat necessary to raise the temperature of 1 gram (1 cm^3) of water 1 degree Celsius
1 quadrillion Btu = (approximately) 1 exajoule
1 joule = 0.239 calorie = 2.778×10^{-7} kilowatt-hour

Mass and Weight

1 pound = 453.6 grams = 0.4536 kilogram = 16 ounces
1 gram = 0.0353 ounce = 0.0022 pound
1 short ton = 2,000 pounds = 907.2 kilograms
1 long ton = 2,240 pounds = 1,008 kilograms
1 metric ton = 2,205 pounds = 1,000 kilograms
1 kilogram = 2.205 pounds

PEARSON ALWAYS LEARNING

Environmental Science II Lab Manual

Custom Edition for Lone Star College

Taken from:
Environmental Issues: Looking Towards a Sustainable Future, Fourth Edition
by Robert L. McConnell and Daniel C. Abel

Ecology on Campus: Lab Manual
by Robert W. Kingsolver

Cover Art: Courtesy of Stockbyte/Getty Images, Corbis.

Taken from:

Environmental Issues: Looking Towards a Sustainable Future, Fourth Edition
by Robert L. McConnell and Daniel C. Abel

Upper Saddle River, New Jersey 07458

Ecology on Campus: Lab Manual
by Robert W. Kingsolver

1900 E. Lake Ave., Glenview, IL 60025

This special edition published in cooperation with Pearson Learning Solutions.

Pearson Learning Solutions, 501 Boylston Street, Suite 900, Boston, MA 02116
A Pearson Education Company
www.pearsoned.com

Printed in the United States of America

1 2 3 4 5 6 7 8 9 10 V092 17 16 15 14

000200010271882364

SK

ISBN 10: 1-269-88743-2
ISBN 13: 978-1-269-88743-4

CONTENTS

Chapter 6 taken from *Ecology on Campus: Lab Manual* by Robert W. Kingsolver

PREFACE

To the Instructor

The idea for this book arose when we were colleagues at the University of Mary Washington. We grew impatient with a teaching style centered on the faculty member as "lecturer" and expert and the student as "scribe" and novice. We felt that such an approach encourages students to be passive rather than active learners and leads to an unhealthy dependency on the faculty person as "expert."

We also found that many students were afraid of math, were rusty in its use, or were superficially trained in arcane fields of calculus. This lack of math skills often leaves students unprepared to deal with the complexity of today's environmental issues. Moreover, we are continually surprised to discover how many bright students can't do three things: understand and USE the units of the metric system, use scientific notation, and critically evaluate complex environmental issues.

Most of a student's discomfort with math is generally founded in frustration. For example, making one error in a series of calculations can render the whole effort useless. We believe that, in the absence of a real learning disability, to solve most math problems requires no special aptitude, only clear sequential instructions and attention to detail. That is why step-by-step calculations are included in your Answer Key.

One of our major objectives is thus to help develop math literacy (*numeracy*) among today's students. We understand that many students have some "math anxiety," so we have included a section entitled "Using Math in Environmental Issues," in which we use a step-by-step method to take students through examples of the calculations in the chapters. We even show sample keystrokes involved in using common calculators. We believe this method will gradually build the student's confidence enough to trust his or her own efforts. Math proficiency is one of the important skills necessary for fully understanding environmental issues, and without these skills, the student's only option is to make choices on the basis of which "expert" is most "believable." Such skills involve the ability to manipulate large numbers using scientific notation and exponents, the ability to use compound growth equations containing natural logs, and so on.

It is our goal that this book be provocative, factually accurate, and up-to-date. *Environmental Issues: An Introduction to Sustainability* is meant be the basis for an issues-oriented introductory, seminar, upper-level, or laboratory course in environmental science or studies. In addition, this newly revised fourth edition is appropriate as a stand-alone text for courses in sustainability or sustainable development. However, it can also be used as a supplement to traditional texts in environmental science, geology, biology, and other natural sciences. It is also suitable for humanities courses that seek to cultivate an awareness of and knowledge about environmental issues and sustainability. You can use as projects to develop students' critical thinking skills in a deliberate and structured way. By their nature, they require students to integrate topics from across subdisciplines to measure, analyze, and evaluate each issue using the discipline and method of a scientist.

But becoming educated is much more than simply acquiring skills. Therefore, we have two additional objectives: to provide students with the knowledge and intellectual standards necessary to apply critical thinking to environmental studies and sustainability, and to foster their ability to critically evaluate issues.

As such, *Environmental Issues: An Introduction to Sustainability* is as much an *interactive* workbook as a traditional textbook. We expect students to have access to standard references in environmental, physical, and natural sciences and to have access to and know how to use the Internet. Indeed, every chapter contains URLs (uniform resource locators) to websites for up-to-the-minute information.

We also trust that you, as an expert in your field and with your own perspectives, will supplement the information in this book with your own comments, introductions, and critical comments on the questions we ask your students. The issues are user-friendly and based on solid science. We define key terms, and keep jargon to a minimum. When important terms are introduced they are italicized and defined if necessary. We introduce and explain key mathematical formulas using a step-by-step nonthreatening approach that we hope you will appreciate.

As we have mentioned, one of our major objectives is to foster proficiency among today's students in the kind of math they need to properly quantify environmental issues, such as the use of key formulas, scientific notation, and the metric system. We provide detailed introductions for each of these topics, as well as a detailed Answer Key for you to use as you see fit that shows the step-by-step calculations used to determine the answers. We have purposefully not provided direct student access to the Answer Key. In addition, we provide you with suggested answers to the "Critical Thinking" questions, but we are confident that you will have your own point of view that you will wish to develop for many if not most of them.

To encourage rigorous critical thinking, there are questions with spaces for answers integrated into each issue. Critical thinking implies using a set of criteria and standards by which the reasoner constantly assesses his or her thinking. At the core of critical thinking is self-assessment.

We devote a detailed section to "critical thinking" in the first section of this book. We think *it is vitally important that students read this material for content, as they will be asked to apply these standards and criteria throughout the book.*

Many questions have been designed to be provocative. This may lead you or your students to perceive a "bias" in the wording of some of the questions. Although we have made the content as factual as possible, we do have strong convictions about these issues. *Convictions are not, however, biases.* Based on the scientific method, our views as scientists are subject to change as evidence supporting our convictions changes. And as such, we are constantly testing the *assumptions* that we use when approaching complex environmental issues. Indeed, this aspect can be turned to a major advantage. Ask your students to look for examples of bias in the questions, and then discuss with them the difference in science between "bias" and "conviction." No doubt it will prove a fruitful activity and may lead students into research (perhaps to "prove us wrong"), which is the essence of progress in the search for scientific truth.

How to Use This Book Effectively

Here's how we have used *Environmental Issues* in our classes. First, we typically dedicate class time or a portion of a laboratory early in the semester to introducing students to the principles of critical thinking. In this period we ask students to list characteristics of critical thinking (i.e., higher order thinking, or just plain good, effective thinking). More often than not, the class's list encompasses many of the standards of critical thinking that are contained in the section of this book entitled Basic Concepts and Tools: Using Math and Critical Thinking. As we go over these characteristics, we emphasize *clarity, awareness of assumptions,* and *continuous self-assessment*, as well as the importance of applying critical thinking to environmental (and other) issues. We then analyze passages from letters to the editor, newspaper op-eds, and popular magazine articles.

It is also worthwhile, before using this book, to confront students' math anxiety early and attempt to reassure them that they are capable of doing the math, although they may be rusty and require some practice and assistance.

We frequently assign chapters as group projects, in which students collaborate using a webpage, "blackboard," or other form of electronic discussion group, and we try to ensure that every student logs on to the system. They can exchange information, send reports, references, and less-threatening critiques to classmates, and so on. We recommend you encourage students to do this, while reminding them that few people are won over to another's argument by having their own ideas ridiculed.

For classes of more than twenty-four students, use of a group e-mail system or other similar classroom software package offers you enormous possibilities. You can break the class into working groups of three to four members, and the group can be reorganized as the term progresses. Working groups can do the calculations independently, can check their work by e-mail or in meetings, and can get together to hash out the answers to the "For Further Thought" questions.

This method also allows you to monitor the class's work and communicate with the class in a nonthreatening manner. Your institution's computer or information technology services department can provide you with the details if you've never used one.

We have found that many students who would be reluctant to participate in a classroom discussion will willingly contribute in the relative anonymity of an electronic discussion group. Students can (and probably should) copy their math and send it to others over the system, thereby checking each other's work. You can send comments to the groups if you feel they are on the wrong track and encourage them if they are making progress. You can send them questions arising out of their own discussions and respond immediately to their inquiries. They can exchange information on the "For Further Thought" questions, and you can discuss these questions with the students as their work evolves.

Students can debate the issues and grade each other, or they can turn in their work in the normal fashion and be graded on the accuracy of their calculations as well as on the thoroughness of their answers.

Work done on these issues can take the place of one or more exams or quizzes, freeing you for other activities and providing students with a less threatening way to earn class credit than an all-exam format. We believe that students retain more information from work on projects and reports than from cramming for tests. At the end of the term, you could pass out a study sheet detailing specifically what material or questions from the chapters will be covered on the final exam, if you choose to test them on their work in this manner.

You may contact us at rmcconne@umw.edu and dabel@coastal.edu with questions and comments.

To the Student

If you are concerned about and interested in environmental issues but feel you just "can't deal with the math," then this book is for you. As environmental scientists, we care deeply about environmental issues, and we feel that you, as a responsible citizen who will have to make increasingly difficult choices in the years ahead, need to be concerned about them as well. Here are a few of our reasons:

- A STRONG scientific consensus exists that our growing human population is having a measurable and harmful effect on the composition of the planet's atmosphere. The world's leaders continue a dialogue on how to respond to human-induced global climate destabilization, a result of global warming. Even though we can't yet be certain what the scale of these changes will be, evidence suggests that the impact will, on the whole, be negative, and could be catastrophic for hundreds of millions of people crowded into many of the planet's coastal cities.
- Marine scientists are concerned about the very survival of many marine organisms, including some that form the basis of major world fisheries. In fact, entire ecosystems such as coral reefs may be at risk. The ocean's ability to absorb our waste, toxic and otherwise, is certainly limited, and these limits have likely been exceeded.
- Although we in the United States and in a few other areas of the world have made impressive strides in improving or at least slowing the degradation of our air, water, and soil, our relentlessly growing human numbers threaten this progress.
- Growing levels of material consumption in developed countries coupled with less-regulated international commerce (free trade) are placing increased stress on critical ecosystems such as tropical and boreal forests, in turn gravely threatening the planet's species diversity.
- Environmental issues, such as water conflicts in the Middle East and the transnational impact of air and water pollution could destabilize international relations and lead to regional conflicts.
- Fossil fuels continue to be the basis of industrial and postindustrial society. Their extraction, transportation, and use impose significant costs on the planet. Much of this cost is externalized (or dumped) onto the environment. As the 2010 Deepwater Horizon Platform incident in the Gulf of Mexico illustrated, addressing environmental issues should involve an inclusion of these costs.

We hope you will find this book to be a provocative introduction to a number of these issues and to many others that you may have never even thought about. These are real-life issues, not hypothetical ones, and you need certain basic skills to fully understand them.

- You must be familiar with the units of the metric system.
- You must be able to use a few mathematical formulas to quantify the issues you will be debating, and you must be able to carry out the calculations accurately.
- We show you how to do this.
- You should develop the habit of rigorously assessing your thinking, and you should apply certain critical thinking skills and techniques when discussing the implications of your calculations.
- You should realize that critical thinking requires practice, dedication, time, and an open mind.

We use a step-by-step method to take you through many of the calculations in this book. Math proficiency is one of the important skills necessary for fully understanding environmental issues, and without these skills, your only option is to make choices on the basis of which "expert" you believe.

But becoming educated is much more than acquiring skills. Therefore, we have two further objectives: to provide you with the knowledge and intellectual standards necessary to apply critical thinking to environmental studies, and to foster your ability to critically evaluate issues, a skill without which mechanical skills are of limited value. To that end, we have included a section called "Basic Concepts and Tools: Using Math and Critical Thinking," in which we illustrate and describe each criterion for critical thinking, as well as the framework within which these criteria are applied. It is very important that you read this section carefully and do all of the exercises in it *before* you begin the analyses of the issues.

Let's consider an example: growth. The word is used in many societal, economic, demographic, and environmental contexts, including growth of the economy, growth of the population, growth of impervious surfaces, growth of food production, and growth of energy use. Assessing the impact of growth requires an understanding of a few simple equations such as the compound interest equation. You should be able to use it accurately and understand its implications. We show you how.

As our national and global population grows and changes and our relationships with other nations and peoples evolve, environmental issues will become even more complicated. Domestically, demographic and ethnic changes are becoming more important, which in turn requires an enhanced ability to critically evaluate issues.

We don't try to avoid controversial topics such as population growth, personal consumption, the automobile, and immigration.

We hope you will be challenged by the issues discussed in this text and that you will research them and become an "expert" on the topics yourself. In fact, if we may be allowed a hidden agenda, this is it.

And while you are testing your own thinking and reasoning, test ours as well. Look for examples of "bias" in the "Critical Thinking" questions, and be prepared to support your conclusions and discuss them with your fellow students and with your instructor.

A Note on Conventions Used in This Book

When we express rates, concentrations, and so on—for example, milligrams per liter, people per hectare, tonnes per year—we use all acceptable formats, including "per" (as above), the slash (mg/L, people/hectare, tonnes/year), and at times negative exponents ($mg \cdot L^{-1}$, $people \cdot hectare^{-1}$, $tonnes \cdot year^{-1}$).

A Word about Calculators

A handheld calculator can make analyzing environmental issues easier, or it can be a source of great frustration. It all depends on how carefully you put the information into the calculator and how familiar you are with how to use it.

We recommend buying the simplest calculator you can find that will carry out the tasks you need performed. These are typically sold as "scientific calculators." *We also encourage you to take the time to learn how to use the calculator properly.* The calculator should perform the basic data manipulation functions—such as adding, subtracting, multiplying, dividing, determining squares and square roots—and have a single, rather than multiple, memory that is easy to use. Some additional features you should look for include the following:

- Parentheses () keys
- A $\mathbf{y^x}$ key
- A reciprocal (**1/x**) key
- Ability to do simple statistics, including means (**x**) and standard deviations (**S**)
- An **ex** key
- A **LN** (natural log) key
- An **exp** or **EE** key

ACKNOWLEDGMENTS

This book could not have succeeded without the aid of numerous friends and colleagues. We appreciate the invaluable feedback from the following reviewers of previous editions, whose comments and suggestions materially improved the manuscript: Janet Kotash, Moraine Valley Community College; Debra Rowe, Oakland Community College; Ravi Srinivas, University of St. Thomas; Daniel J. Sherman, University of Puget Sound; and Maud Walsh, Louisiana State University. Special thanks to Coastal Carolina University students Lyndsey King and Chelsea Norman, who helped edit the 4th edition of this book.

Robert L. McConnell
Daniel C. Abel

Rather than to travel into the sticky abyss of statistics, it is better to rely on a few data and on the pristine simplicity of elementary mathematics.
—ALBERT BARTLETT

An unexamined life is not worth living.
—SOCRATES

CHAPTER 1
Basic Concepts and Tools: using Math and Critical Thinking

Here we review the units of the metric system and the rules for using scientific notation in math problems. We will also show you how to do more complicated math, such as projecting population based on growth rates, using step-by-step methods. This chapter also deals with *critical thinking*. At first glance many issues seem perplexing and hard to approach. Don't panic; we will show you how to apply intellectual standards within a critical thinking framework.

THE METRIC SYSTEM

The metric system's elegance and utility arise from its simplicity. You probably already know that the metric system is based on powers of ten: 10 millimeters in 1 centimeter, 100 centimeters in 1 meter, and so forth. This key point should be kept in mind.

You must be able to make certain conversions from the English system to the metric system and back. (It is a good idea to mark this page, since you will find it useful to refer to again.) Some common conversions as well as metric prefixes are given in Tables 1-1 through 1-4 below and may also be found inside the front cover of this book.

Here's a useful shortcut. To convert areas, you can use the conversion factors for the units of length and square them. For example, to convert 6 square feet to square yards, do the following:

$$6\ \text{ft}^2 \times (1\ \text{yd}/3\ \text{ft})^2 = 6\ \cancel{\text{ft}^2} \times 1\ \text{yd}^2/9\ \cancel{\text{ft}^2} = 0.67\ \text{yd}^2$$

We will introduce and explain other units of the metric system as the need arises. One relationship you will find especially helpful: There are 1,000 liters per cubic meter of water under standard conditions.

Now do these self-assessment questions to test your ability to manipulate and convert these units. These skills are essential for the work in the issues that follow. Remember to use your conversion factors (so that units cancel each other out). For example, to determine how many liters are in 3.6 cubic meters:

$$3.6\ \cancel{\text{m}^3} \times 1000\ \text{L}/\cancel{\text{m}^3} = 3600\ \text{L}$$

TABLE 1-1 ■ Units of the Metric System

Units of Distance: The fundamental unit is the *meter*	
1000 (10^3) m = 1 km	One thousand meters = one kilometer
100 (10^2) cm = 1 m	One hundred centimeters = one meter
10 (10^1) mm = 1 cm	Ten millimeters = one centimeter
1000 (10^3) μm = 1 mm	One thousand micrometers = one millimeter
Units of Mass: The fundamental unit is the *gram*	
1000 kg = 1 metric ton	One thousand kilograms = one (metric) tonne
1000 g = 1 kg	One thousand grams = one kilogram
1000 mg = 1 g	One thousand milligrams = one gram
1,000,000 (10^6) ng = 1 mg	One million nanograms = one milligram
Units of Volume: The fundamental unit is the *liter*	
1000 L = 1 m^3	One thousand liters = 1 cubic meter
1000 ml = 1 L	One thousand milliliters = 1 liter

TABLE 1-2 ■ Metric Prefixes and Equivalents

Large Numbers		
One thousand	=1000	$=10^3$ (kilo or k)
One million	=1,000,000	$=10^6$ (mega or M)
One billion	=1,000,000,000	$=10^9$
One trillion	=1,000,000,000,000	$=10^{12}$
One quadrillion	=1,000,000,000,000,000	$=10^{15}$ (commonly used in expressions of energy use)
Small Numbers		
One hundredth	=1/100	$=10^{-2}$ (centi or c)
One thousandth	=1/1000	$=10^{-3}$ (milli or m)
One millionth	=1/1,000,000	$=10^{-6}$ (micro or mc or μ)
One billionth	=1/1,000,000,000	$=10^{-9}$ (nano or n)

TABLE 1-3 ■ Some Common Metric Conversions

gallons/liters	1 U.S. gal. = 3.8 L	One U.S. gallon = 3.8 liters
liters/gallons	1 L = 0.264 U.S. gal.	One litre = 0.264 U.S. gallon
meters/yards	1 m = 1.094 yd	One meter = 1.094 yards
yards/meters	1 yd = 0.914 m	One yard = 0.914 meter
grams/ounces	1 g = 0.035 oz	One gram = 0.035 ounce
ounces/grams	1 oz = 28.35 g	One ounce = 28.35 grams
kilograms/pounds	1 kg = 2.2 lb	One kilogram = 2.2 pounds
pounds/grams	1 lb = 454 g	One pound = 454 grams
miles/kilometers	1 mi = 1.609 km	One mile = 1.609 kilometers
kilometers/miles	1 km = 0.621 mi	One kilometer = 0.621 mile

TABLE 1-4 ■ Conversion Factors for Area

square miles/square kilometers	1 mi^2 = 2.6 km^2	One square mile = 2.6 square kilometers
square kilometers/square miles	1 km^2 = 0.39 mi^2	One square kilometer = 0.39 square mile
hectares/acres	1 ha = 2.47 acres	One hectare = 2.47 acres
acres/hectares	1 acre = 0.4 ha	One acre = 0.4 hectare
square yards/square meters	1 yd^2 = 0.84 m^2	One square yard = 0.84 square meter
square meters/square yards	1 m^2 = 1.2 yd^2	One square meter = 1.2 square yards
square miles/acres	1 mi^2 = 640 acres	One square mile = 640 acres

Question 1-1: How many micrometers are in a meter?

Question 1-2: How many centimeters are in a kilometer?

Question 1-3: How many grams are in a tonne? (*Tonne* is the correct spelling for the metric unit of 1000 kg.)

Question 1-4: Express your height in feet, meters, and centimeters.

Question 1-5: Express your weight in kilograms and pounds.

SCIENTIFIC NOTATION

Very large numbers are most conveniently manipulated (i.e., added, subtracted, multiplied, and divided) by converting the numbers to logarithms to the base 10. Since this is not a math book, we are not going to delve into the theory of logarithms; a practical application is all you need.

The basic fact you need to know is that 100 (pronounced "ten to the zero" or "ten to the zero power") is defined as 1.

The skills to do the manipulations are easy to learn. The first step is to convert large numbers to scientific notation, with which you should already be familiar. For example, in scientific notation, 18,000,000 is 1.8×10^7.

Try another example.

Question 1-6: Express one billion (1,000,000,000) in scientific notation.

Now let's introduce a wrinkle.

Question 1-7: Express 2,360,000 in scientific notation.

You can express the same number in a variety of ways using exponents, such as 23.6×10^5, and they all will mean the same thing. But it is customary to express all values in the same format, typically by placing only one digit to the left of the decimal place (e.g., 2.36×10^6).

Question 1-8: Express 23,000,000,000,000 (23 trillion) the customary way using exponents.

For numbers summarized in scientific notation, the prefix in front of the unit (e.g., kilo) denotes the magnitude of the unit (e.g., *kilo*grams = units of 1000 g).

Question 1-9: Convert 1.86 mm to (a) nm, (b) mm (micrometers), (c) cm, (d) m, and (e) km. Express your answers as decimals and in scientific notation.

MANIPULATING NUMBERS EXPRESSED IN SCIENTIFIC NOTATION

To add, subtract, multiply, and divide large numbers using scientific notation, you need to remember a few basic rules and check your work carefully. That is all there is to it. Here are the rules.

Multiplication Using Scientific Notation

To multiply numbers expressed in scientific notation, *multiply the bases and add the exponents*. For example, to multiply

$$(3 \times 10^3) \times (4 \times 10^5)$$

multiply the bases

$$3 \times 4 = 12$$

and add the exponents

$$3 + 5 = 8$$

The result is 12×10^8 or, using the appropriate convention, 1.2×10^9. (NOTE: This is the same as 120×10^7 and any number of other variations as well.)

Division Using Scientific Notation

To divide numbers expressed in scientific notation, *divide the number in the numerator by the number in the denominator and subtract the exponent of the denominator from the exponent of the numerator*. Recall that:

$$\frac{\text{NUMERATOR}}{\text{DENOMINATOR}}$$

For example, to divide

$$(5.2 \times 10^4) \div (2.6 \times 10^2)$$

divide the numerator by the denominator

$$5.2 \div 2.6 = 2$$

and subtract the exponent of the denominator from the exponent of the numerator

$$4 - 2 = 2$$

So the answer is 2×10^2.

Question 1-10: Perform the following manipulations:

$$(8.7 \times 10^{-3}) \times (4.2 \times 10^{-9}) = \underline{\qquad\qquad}$$
$$(5.2 \times 10^{18}) \times (8.7 \times 10^{22}) = \underline{\qquad\qquad}$$
$$(8.7 \times 10^{-3}) \div (4.2 \times 10^{-9}) = \underline{\qquad\qquad}$$
$$(5.2 \times 10^{18}) \div (8.7 \times 10^{22}) = \underline{\qquad\qquad}$$

Addition Using Scientific Notation

To add numbers expressed in scientific notation, *you simply add the numbers after converting both to the same exponent.*

For example, to add 3 billion to 14 million,

$$3{,}000{,}000{,}000 + 14{,}000{,}000$$

or

$$(3 \times 10^9) + (14 \times 10^6)$$

convert both numbers to the same exponent (it doesn't matter which, but it is usually easier to use the smaller):

$$3 \text{ billion} = 3000 \text{ million or } 3000 \times 10^6$$
$$14 \text{ million} = 14 \times 10^6$$

Therefore,

$$(3000 \times 10^6) + (14 \times 10^6) = 3014 \times 10^6$$

An equally correct answer would be 3.014×10^9 which is the customary way to express the answer.

Now work out the rule for subtraction using scientific notation.

USING MATH IN ENVIRONMENTAL ISSUES

The following is an introduction to some of the formulas used in this book.

How to Project Population Growth Using the Compound Growth Equation

For this you will need a calculator with an exponent key. The equation is

future value = present (or starting) value × (e)^{rt}

where e equals the constant 2.71828 . . . , r equals the rate of increase (expressed as a decimal, e.g., 5% would be 0.05), and t is the number of years over which the growth is to be measured.

Replacing words with symbols, this equation becomes

$$\mathbf{N = N_0 \times (e)^{rt}}$$

The variable N_0 represents the value of the quantity at time zero, that is, the starting point.

This equation is central to understanding exponential growth and is one of the few worth memorizing. Using it is not as intimidating as you may think.

Sample Growth Calculation. Here's an example. Let's figure out (demographers use the word *project*) the world population in 2020, given the mid-year 2006 population of 6.52 billion (6.52×10^9 or 6,520,000,000) and a growth rate of 1.14% per year. You can obtain current monthly world population figures at the Census Bureau's World POPClock website (www.census.gov/cgi-bin/ipc/popclockw).

Here's how to do this calculation:

$$2020 \text{ population} = (6.52 \times 10^9) \times e^{(0.0114 \times 14)}$$

On a typical, nongraphics calculator, keystrokes are as follows (commas are for punctuation only):

Key in **0.0114** (the decimal equivalent of 1.14%), then × (multiply sign), then **14,** then **=**. This gives you the exponent (the number to which e must be raised). Next hit the button labeled $\mathbf{e^x}$. Note that on most calculators, $\mathbf{e^x}$ is labeled *above* a button having another label (frequently **ln**). If this is the case on your calculator, you must hit the key labeled **2nd** or

2nd F first, followed by the key with $\mathbf{e^x}$ above it. Further, some calculators require you to hit the = key after the $\mathbf{e^x}$ key.

Next, key in × (multiply sign), followed by the *starting value* (6.52×10^9). On most calculators this is done by keying in **6.52,** then hitting the button labeled **EE** or **EXP,** then keying in **9.** If the **EE** or **EXP** label is above the key, recall that you must first hit the key labeled **2nd** or **2nd** F before punching **EE** or **EXP**. Finally, hit the = sign.

The correct answer is 7.65 billion.

Now, project the population of Dhaka, Bangladesh, in 2056 at a growth rate of 4.2% per year using the compound growth equation. The 2005 population was 12.6 million.

$$\begin{aligned} \textbf{2056 population} &= (12.6 \times 10^6) \times e^{(0.042 \times 51)} \\ &= 1.07 \times 10^8 \\ &= 107 \times 10^6 \\ &= 107 \text{ million} \end{aligned}$$

The compound growth equation can also be rearranged. If you know the starting and ending population sizes over a given period, you can calculate the average growth rate over that period using the formula

$$r = (1/t) \ln(N/N_0)$$

Also, you can calculate how long it would take a population of a given size to grow (or decrease) to a different size at a specified growth rate using

$$t = (1/r) \ln(N/N_0)$$

Doubling Time

When a population grows exponentially (by a percentage of the original number), the time it takes for the population to double, called *doubling time* (symbol "**t**"), can be approximately calculated using the following formula:

$$t = (70/r)$$

where t is the doubling time (usually in years) and r is the growth rate expressed as the decimal increase or decrease × 100 (for example, you would enter 7 for a 7% increase).

Derivation of Doubling Time. To derive the doubling time formula, we revisit the compound growth equation:

$$\textbf{future value} = \textbf{present value} \times (e)^{rt}$$

where e equals the constant 2.71828..., r equals the rate of increase (expressed as a decimal, i.e., 5% would be 0.05), and t is the number of years (or hours, days, etc.—whatever units you are using in r) over which the growth is to be measured.

Replacing words with symbols, this equation becomes:

$$N = N_0 \times (e)^{rt}$$

The variable N_0 represents the value of the quantity at time zero, that is, the starting point.

If a population doubles in size (that is, increases by a factor of 2), the ratio N/N_0 would be exactly 2.

The equation is thus rearranged as follows:

$$N/N_0 = e^{rt}$$

$$2 = e^{rt}$$

Taking the natural log of each side of the equation, we get

$$\ln 2 = \ln(e^{rt})$$

$$\ln 2 = rt$$

$$\ln 2 = 0.693$$

$$0.693 = rt$$

Dividing by r:

$$0.693/r = t$$

For convenience sake, we round 0.693 to 0.70, and we also multiply the left side of the equation by 100/100, which allows us to enter the rate as a percentage (i.e., 5% would now be entered as 5 instead of 0.05). Thus the final doubling time formula:

$$t = 70/r$$

CRITICAL THINKING[1]

> "It is a humanist principle that if you want to know the truth, go to the sources, not to the commentators."[2]

Overview

The mind has awesome power. John Milton said,

> The mind is its own place,
> And in itself can make
> A Heav'n of Hell,
> A Hell of Heav'n.

We must use the mind's power effectively, which involves critical thinking. Critical thinking is sometimes called *second order thinking. First order thinking* is spontaneous, often emotional, and rarely analytical and reflective. As such, it contains prejudice, bias, truth and error, inspiration, and distortions; in short, good and bad reasoning, all mixed together. Second order thinking is essentially first order thinking "raised to the level of conscious realization," that is, analyzed, assessed, and thereby reconstructed.[3]

---

[1] We obtained the basis for much of the information in this section from handouts, discussions, and workshops at the 12th to 20th International Conferences on Critical Thinking and Educational Reform sponsored by Sonoma State University's Center for Critical Thinking and Moral Critique.

[2] Barzun, J. 2000. *From Dawn to Decadence: 1500 to the Present* (New York: HarperCollins), p. 54.

[3] Paul, R., & L. Elder. 2000. *Critical thinking—Tools for taking charge of your learning and your life* (Upper Saddle River, NJ: Prentice Hall).

Many scientists equate critical thinking with the application of the scientific method, but we think critical thinking is a far broader and more complex process. Critical thinking involves developing skills that enable you to dissect an issue (*analyze*) and put it together (*synthesize*) so that interrelationships become apparent. It involves searching for *assumptions,* the basic ideas and concepts that guide our thoughts. Critical thinking also encourages an appreciation for our own and for others' *points of view,* which is important when approaching complex environmental issues.

Too often, analyzing complex issues leads some to a belief that everyone is "entitled" to an opinion that should be respected. We do not necessarily concur. However, problem solving demands a willingness to listen for *content* to what others are saying. *Talking is easy, but listening is not.* Developing critical thinking skills is not like learning to ride a bicycle. All of us must learn to use a set of intellectual standards as an "inner voice" by which we constantly test and hone our reasoning skills. But the standards must be set in an appropriate framework in order for true critical assessment to take place.

The following paragraphs describe the intellectual standards you should apply when assessing the quality of your reasoning. This is the basis for critical thinking, which in turn is the approach that we try to apply throughout this book.

Intellectual Standards: The Criteria of Solid Reasoning

Clarity. This is the most important standard of critical thinking. If a statement is not clear, its accuracy or relevance cannot be assessed. For example, consider the following two questions:

1. What can we do about global climate change?
2. What can citizens, regulators, and policy makers do to make sure that greenhouse gas emissions from industry, transportation, and power generation do not cause irreversible ecological damage or harm human health?

Accuracy. Is the statement true? How can we find out? A statement can be *clear* but not *accurate*.

Precision. Can we have more details? Can you be more specific? A statement can be clear and accurate, but not precise. For example, we could say that "There were more sport utility vehicles (SUVs) in the United States in 2012 than in 2001." That statement is clear, it is accurate, but how many more SUVs are there? 1? 1,000? 1,000,000? (Note that there is a difference between the way many scientists use the word *precision* and the more general way it is used here.)

Relevance. How is the statement or evidence related to the issue we are discussing? A statement can be clear, accurate, and precise, but not relevant. Here's an example: If we are given the responsibility to eliminate the harmful environmental impact of pollutants from coal-burning power plants, and we invite public comment on our proposals, someone might say, "Electricity from coal-burning power plants accounts for 100,000 jobs in this state alone." That statement may be clear, accurate, and precise, but it is not relevant to the specific issue of removing pollution (although in other contexts it may in fact be very relevant).

Here is another example: Parks in Arlington, Virginia, have signs at all entrances that read "Dogs must be on leash and under control at all times" and include the relevant County Code citation. Nevertheless, dog owners typically ignore the signs. Here are some of their reasons: "The neighbors don't complain," "There is no dog exercise area near my home,"

"My dog is well-behaved and doesn't need a leash," and "The police don't mind if our dogs play here." Using critical thinking, assess the relevance of the dog owners' responses.

Breadth. Is there another point of view or line of evidence that could provide us with some additional insight? Is there another way to look at this question? For example, you will assess the issue of turf grass proliferation in Chapter 8. In an article one of us (RLM) wrote for the *Washington Post,* he suggested that since turf-care devices such as mowers, trimmers, blowers, and the like are significant sources of air pollution, and since their use is proliferating, it might be easier to address the problem they pose not by banning or overregulating these devices, but by reducing the area of turf that must be maintained.

Depth. How does a proposed solution address the real complexities of an issue? Is the solution realistic or superficial? This question is one of the most difficult to tackle, because here is where reasoning, "instinct," and moral values may interact. The points of view of all who take part in the debate must be carefully considered. For example, politicians have suggested "Just don't do it" as a solution to the problem of teenage drug use, including smoking. Is that a realistic solution to the problem, or is it a superficial approach? How would you defend your answer? Is your defense grounded in critical thinking?

Logic. How does one's conclusion follow from the evidence? Does the conclusion really make sense? Why or why not? When a series of statements or thoughts are mutually reinforcing and make sense together, and when they exhibit the intellectual standards described above, we say they are logical. When the combination does not make sense, is internally contradictory, or not mutually reinforcing, it is not "logical." Logic in an argument is to the trained mind a bit like the apocryphal definition of obscenity: "You know it when you see it" (see the following section on "Logical Fallacies and Critical Thinking").

Applying Intellectual Standards in a Critical Thinking Framework

The intellectual standards described above are essential to critical evaluation of environmental issues, but there are more factors to be considered. The following criteria constitute the framework in which these standards should be applied.

- **Point of view.** What viewpoint does each contributor bring to the debate? Is it likely that someone who has a job in a weapons plant would have the same view on military spending as someone who doesn't? Would a tobacco company executive be likely to have the same opinion on restricting smoking as someone who lost a relative to lung cancer? Think of other examples, but note that *identifying a point of view does not mean that the opinion should automatically be accepted or discounted.* We should strive to identify our own point of view and the bases for this, we should seek other viewpoints and evaluate their relevance, and we should strive to be fair-minded in our assessment. Few people are won over by having their opinions ridiculed. It is important to note here that our points of view are often informed by our *assumptions*, which we will address below.
- **Evidence.** Scientific problem solving is, or should be, based on evidence and information (sometimes called *data*, but we prefer to apply this term solely for numbers used in calculations). Our conclusions or claims must be based on sufficient relevant evidence, the information must be laid out clearly, the evidence against our position must be evaluated, and we must be open to new evidence that challenges our conclusions.
- **Purpose.** All thinking to solve problems has a purpose. It is important to have a clear understanding of that purpose and to ensure that all participants are on the "same wavelength." Since it is easy to wander off the subject, it is advisable to periodically check to make sure the discussion is still on target. For example, stu-

dents working on a research or term project or employees tackling a work-related problem occasionally stray into subjects that are irrelevant and unrelated, although they may be interesting or even seductive. It is vitally important, therefore, that the issue being addressed must be defined and understood as precisely as possible.

- **Assumptions.** Here is an excerpt from a January 2001 report prepared by the U.S. Energy Information Administration: "With a growing economy, U.S. energy demand is projected to increase 32 percent from 1999 to 2020, reaching 127 quadrillion BTU, *assuming* no changes in Federal laws and regulations." This statement is clear, precise, and contains assumptions.

  All reasoning and problem solving depends on *assumptions*, which are *statements accepted as true without proof.* When we assume something, we presuppose it or take it for granted. For example, students show up in class because they assume their professor/teacher will be there. "Never assume" is an old adage. However, it is more reasonable to be aware of and take care in our assumptions and always be ready to examine and evaluate them. They often need to be revised in the light of new evidence.

Now, before we analyze our own assumptions, let's summarize some characteristics of sound reasoning.

Skilled reasoners

- Understand key concepts (such as externalities) and ideas.
- Can explain key words and phrases (such as global warming).
- Can explain scientific terms (such as bioaccumulation).
- Continually exercise their thinking skills.
- Can recognize irrelevant topics and can explain why they are irrelevant.
- Come to well-reasoned conclusions and solutions.

Additionally, the effective reasoner continually assesses and reassesses the quality of his or her thinking in light of new evidence. Finally, the effective reasoner must be able to communicate effectively with others.

LOGICAL FALLACIES AND CRITICAL THINKING

After you leave this course, much of your information concerning environmental issues will come from the media—television, magazines, radio talk shows, and newspapers. These sources often exhibit evidence of poor reasoning—that is, logical fallacies. Learn to identify them to ensure you are getting the best information possible. Here, in no particular order, are some of the more common examples.

- The fallacy of *composition*: assuming what is good for an individual is good for a group. An example is standing at sports events—an advantage to one person but not when everybody does it.
- The fallacy of *starting with the answer*: including your conclusion in your assumptions. For example, "America runs on high levels of energy consumption. Without energy, we wouldn't have the American lifestyle. We can't have the American lifestyle without energy. Thus, America can't afford to conserve energy." Here, the arguer is simply defining his way out of the problem. By his reasoning, we would have to continue to increase energy consumption forever, an obvious impossibility.
- The fallacy of *hasty generalization*: "Senegal and Mali have very low levels of energy consumption. They are very poor countries. Low levels of energy consumption lead to poverty." How is poverty measured? Are there any "wealthy" countries that have relatively low levels of energy consumption? Are there any relatively poor countries that have high levels of energy consumption?

- The fallacy of *false choice*: stating an issue as a simplistic "either-or" choice when there are other more logical possibilities. "Those who don't support fossil fuel use want to go back to living in caves."
- Fallacy of an *appeal to deference*: asserting an argument because someone famous supports it.
- Fallacy of *ad hominem* (literally, "at the person") argument: attacking a person or a person's motives without discussing the merits of his or her position.
- The fallacy of *repetition*—the basis of most advertising: repeating a statement over and over without offering any evidence. "Population growth is good. People contribute to society. We need population growth to survive."
- The fallacy of *appealing to tradition*: "Coal built this country. Reducing coal use would threaten our society."
- The fallacy of *appealing to pity*: "Three million people work in the mining industry. Regardless of its impact, we have to support them."
- The fallacy of an *appeal to popularity*: "Seventy-five percent of Americans support this position." Perhaps the poll asked the wrong question; perhaps the respondents didn't have enough information to properly respond, etc.
- The fallacy of *confusing coincidence with causality*: "After passage of the Endangered Species Act (ESA), jobs in sawmills fell 70%. Therefore the ESA was bad for the economy." Were there other possible explanations for the drop in jobs?
- The fallacy of the *rigid rule*: "Hard-working people are good for the economy. Immigrants are hard-working people. Therefore, the more immigrants we have, the better for our economy." Or, "Large numbers of immigrants commit crimes. Crimes are bad for the economy. Immigration should be ended."
- The fallacy of *irrelevant conclusion*: using unrelated evidence or premises to support a conclusion. "Development raises the value of land. Development provides jobs. Developed land pays more taxes than undeveloped land. Therefore flood plain land should be developed."

Assumptions about Environmental Issues

We cannot stress too heavily the power of assumptions in guiding our reasoning. To give you an example, in the following passages we would like you to respond to the following real-world issues:

First, identify your assumptions in defining government's role in protecting the environment.

Second, identify your assumptions as to the proper *level* of government that may act.

Third, determine your position on the "precautionary principle."

Fourth, identify your assumptions concerning the extent to which individuals or institutions in a society may impose costs upon others in that society with or without their knowledge or consent.

THE ROLE OF GOVERNMENT

Respond to the following quote, taken from Thomas Jefferson's First Inaugural Address, delivered on March 4, 1801. Most politicians and most Americans probably consider themselves to have "Jeffersonian" principles.[4] Here is what Jefferson said:

> What more is necessary to make us a happy and prosperous people? Still one thing more, fellow citizens—a wise and frugal Government, <u>which shall restrain men from injuring one another,</u>

[4] See for example www.lewrockwell.com/vance/vance17.html.

shall leave them otherwise free to regulate their own pursuits of industry and improvements, and shall not take from the mouth of labor the bread it has earned.[5]

Question 1-11: In a clear sentence or two, explain what you think Jefferson meant by the phrase we underlined.

Question 1-12: Do you think he was referring solely to thugs who physically brutalize their fellow citizens? Explain.

Question 1-13: Could he logically have been referring also to citizens who sought to poison others? In other words, is restraining poisoners a legitimate role of government? Explain.

Question 1-14: Now, what if a citizen or an organization dumps a toxin into water or air that all citizens depend on, or if a citizen or an organization fills in a wetland that performed valuable ecological functions upon which local residents depend? May government under Jefferson's principle restrain that person or organization?

Your answer to these questions will define your assumptions as to the proper role of government.

The Proper Level of Government That May Act

Next, we will ask you to evaluate your assumptions about the *level* of government that may properly intervene in environmental issues.

One of the major discoveries of the past two decades has been the extent to which much pollution is *transboundary* in nature. For example, as much as one-third of the NOx (oxides of nitrogen) air pollution affecting the Washington, DC, metropolitan area annually comes from as far away as the Midwest and southern Canada; much of the pollution

[5] See http://avalon.law.yale.edu/subject_menus/inaug.asp.

that degrades air quality over Grand Canyon National Park comes from southern California, several hundred kilometers to the west; some of Oregon's air pollution comes from coal-fired electric utilities in China; and so forth.

Question 1-15: Therefore, is it appropriate that local government primarily or solely bear the responsibility for protecting its own environment? May the states and federal government have a legitimate role based on the transboundary nature of pollution? Explain and justify your answer, using additional paper if necessary.

Your answer will help evaluate your assumptions about the extent to which state and federal government agencies have responsibilities to intervene to protect local environments. Remember to reassess your assumptions in the light of new evidence.

The Precautionary Principle

Scientists generally define the *precautionary principle* as follows:

> Action should be taken to prevent damage to the environment even in cases where there is no absolute proof of a causal link between emissions or activity and detrimental environmental effect. Embedded in this is the notion that there should be a reversal of the "burden of proof" whereby the onus is now on the operator to prove that his action will not cause harm rather than on the environment to prove that harm (is occurring or) will occur. [6]

Another way to express this principle is "better safe than sorry." Many products of science and technology are brought to the marketplace without adequate knowledge of possible long-term effects on human health and the global environment. Some examples are the uses of freon, mercury, and organochlorines, which you will investigate later.

In most industrialized nations, the so-called "burden of proof" falls not on the producers of goods but rather on those who allege that they have suffered harm. This is the basis of our tort system of civil law. As a result of the proliferation of new products, government agencies like the Food and Drug Administration, the Environmental Protection Agency (EPA), and the Federal Trade Commission, to name but a few, are sometimes unable to keep pace. For example, as of 2010, California had registered more than 900 pesticide active ingredients that were used in approximately 12,000 pesticide products. One of six boards and departments within Cal/EPA, the Department of Pesticide Regulation, regulates the sale and use of pesticides to protect human health and the environment.

Although individuals have recourse to law if they believe they have been injured, future generations, wildlife, and ecosystems have no such means of redress. Adherence to the precautionary principle could in the view of many facilitate democratic oversight.

Similarly, a serious threat like global warming or the proliferation and buildup of organochlorines under the precautionary principle would trigger action to address the threat, even if the "science" is not yet conclusive but is supported by the preponderance of available evidence.[7]

---

[6] Glegg, G., & P. Johnston. 1994. The Policy Implications of Effluent Complexity. In *Proceedings of the Second International Conference on Environmental Pollution* (London: European Centre for Pollution Research), Vol. 1, p. 126.

[7] Shabekoff, P. 2000. *Earth Rising: American Environmentalism in the 21st Century* (Washington, DC: Island Press).

Question 1-16: Discuss your opinion on the precautionary principle. Should those who wish to introduce a new chemical, a new industrial process, a land-use change, and so on, have to demonstrate that their change will not harm the environment before proceeding? Explain and defend your answer, using additional paper if necessary.

The Question of Externalities

Economists define *externalities* as any *cost* of production not included in the *price* of the good. An example would be environmental pollution or health costs resulting from burning diesel fuel not included in the price of the fuel.[8] Another example would be cleanup costs paid by governments resulting from animal waste degradation of water bodies from large-scale meat processing operations. In this example, the price of chicken or pork at your local supermarket is lower than it would be if all environmental cleanup costs were included in the price of the meat.[9]

Question 1-17: Consult an economics textbook or do a search on the Internet using the term externalities. State whether you conclude that externalities should be included in the costs of goods, or whether and in what circumstances some costs can be left for others to pay. Justify your answer using those principles of critical thinking outlined previously.

Assumptions about Corporations

Here is another quote from Thomas Jefferson, on the impact of those new organizations called corporations. Read Jefferson's words and then respond to the following question.

> I hope we shall take warning from, and example of, England, and crush in its birth the aristocracy of our moneyed corporations, which dare already to challenge our Government to trial, and bid defiance to the laws of our country.

Question 1-18: Do you share or reject Jefferson's opinions concerning corporations? Cite evidence or provide support to your conclusion. It might help you to prepare a list of positive and negative contributions corporations make to our economy. Do you feel corporations have too much power in contemporary life? Why or why not? Use additional paper if necessary.

If you are interested in corporate power, research the 1886 U.S. Supreme Court ruling: *Santa Clara County v. Southern Pacific*. The Court ruled that Southern Pacific was a "natural person" entitled to the protections of the U.S. Constitution's Bill of Rights and Fourteenth Amendment.

---

[8] For background on diesel, see California Air Resources Board, www.arb.ca.gov/homepage.htm.

[9] See for example Environmental Defense www.environmentaldefense.org.

Then review "Citizens United vs Federal Election Commission" a major 2010 decision that held that, under the 1st Amendment to the U.S. Constitution, the federal government cannot limit political contributions from corporations or labor unions. The court did uphold the existing prohibition on donations by these groups to individual political candidates.

After researching these cases, answer these questions:

Question 1-19: Discuss whether you believe the Court acted correctly in deciding that a corporation was a person.

Question 1-20: Is the 1886 ruling relevant to the twenty-first century? Why or why not?

Question 1-21: Do you agree with the "Citizens United" decision (a 5–4 decision by the way)? Why or why not?

After having thoughtfully responded to the above scenarios, you should now have a better awareness of the assumptions that you bring to the analysis of environmental issues that you are about to undertake.

SUMMARY

To summarize, the intellectual standards by which critical thinking is carried out are clarity, accuracy, precision, relevance, breadth, depth, and logic. These standards are applied in a framework delineated by points of view, assumptions, evidence or information, and purpose. We encourage you to return to this section whenever you need to refresh and polish your critical thinking skills.

CHAPTER 2
Principles of Sustainability

On September 22, 2009 President Barack Obama addressed the United Nations. He began his speech: "The threat from climate change is serious, it is urgent and it is growing. Our generation's response to this challenge will be judged by history, for if we fail to meet it—boldly, swiftly and together—we risk consigning future generations to an irreversible catastrophe."[1]

Seventeen years previous, in 1992, 1700 scientists, including the majority of living Nobel laureates in the sciences, issued the World Scientists' Warning to Humanity:

> Human beings and the natural world are on a collision course. Human activities inflict harsh and often irreversible damage on the environment and on critical resources. If not checked, many of our current practices put at serious risk the future that we wish for human society and the plant and animal kingdoms, and may so alter the living world that it will be unable to sustain life in the manner that we know. Fundamental changes are urgent if we are to avoid the collision our present course will bring about.[2]

The solution to these environmental problems is sustainability.

Question 2-1: Why do you think so little was done to address climate change between 1992 and 2009?

WHAT IS *SUSTAINABILITY*?

Sustainability, or sustainable development, was defined in 1987 by the World Commission on Environment and Development as development that meets the needs of the present without compromising the ability of future generations to meet their own needs.[3] Paul Hawken, author of *Natural Capital*, and *The Ecology of Commerce*, defined sustainability more practically: "Leave the world better than you found it, take no more than you need, try not to harm life or the environment, make amends if you do."[4]

Sustainability transcends and supersedes environmentalism. It involves a transformation from a wasteful linear model of resource use in which natural resources are extracted, converted to goods, then trashed or landfilled, to a cyclical model built around efficiency, waste reduction, reuse, and recycling. Moreover, environmentalism focuses primarily on the natural, nonhuman world and the effects humans have on it, whereas sustainability adds social and economic justice components. Thus, sustainability is based on what is known as the *triple bottom line*: planet, people, and prosperity.

---

[1] http://cop15.state.gov.

[2] http://www.ucsusa.org/about/1992-world-scientists.html.

[3] World Commission on Environment and Development (Gro Harlem Brundtland, Chair). 1987. *Our Common Future* (New York: Oxford University Press).

[4] Hawken, P. 1994. *The Ecology of Commerce* (New York: HarperCollins).

MEASURING SUSTAINABILITY

Measuring sustainability is not straightforward. In 2005 the Environmental Performance Measurement Project at Yale University issued an Environmental Sustainability Index (ESI), ranking nations on the extent to which their societies approached sustainability. This Index was updated in 2012.[5] The countries ranked as "strongest performers" were in order: Switzerland, Latvia, Norway, Luxembourg, Costa Rica, France, Austria, Italy, United Kingdom, and Sweden. The U.S. ranked 49th. Iraq ranked last, at 132. In most cases, however, insufficient data exist to accurately determine each nation's ESI. Their rankings are, therefore, only crude comparative measurements of societal sustainability.

The Sustainable Communities Network sets general targets for a Sustainable Planet Earth. They are[6]

Creating Community
Growing a Sustainable Economy
Protecting Natural Resources
Governing Sustainably
Living Sustainably

In our view achieving sustainability requires addressing agriculture, fisheries, forests and wood products, water supplies, energy, biodiversity, climate change, manufacturing and industry, and justice and equity. Each of these is addressed as individual chapters, or within chapters.

Agriculture

A sustainable society must preserve agricultural land, practice sustainable agriculture, and produce substantial food supplies locally. Agriculture should be based ultimately on organic methods. In the United States and European Union (EU), the consumption of organic food has been increasing at 15 to 20 percent annually for more than a decade. Progress toward sustainable agriculture has been slowed by agricultural subsidies, which distort markets and often harm the environment. In 2012 *The Economist* estimated agricultural subsidies from the EU, the U.S., Japan, China, Russia, and Brazil alone at $360 billion.[7]

Eventually, the impact of the growing global demand for meat must be addressed. Contemporary industrial-style meat production imposes significant environmental costs. For example, cattle-raising poses one of the greatest threats to the survival of tropical rainforests. Moreover, per capita yields from global fisheries are declining, and expansion of "aquaculture" is often at the expense of coastal ecosystems and wild fish stocks (see "Fisheries").

Fisheries

Many fish species ranging from sardines to bluefin tuna are in dangerous decline. Thriving and diverse aquatic wildlife are necessary for healthy marine and freshwater ecosystems. It is therefore critical that communities dependent on fisheries and aquatic ecosystems use these resources responsibly. One successful method is creating marine protected zones that are large enough to be self-sustaining, thus protecting species diversity from human interference. But protected zones alone may not be enough: Many large fish and marine

[5] For data on which these rankings were based, go to http://epi.yale.edu/epi2012/methodology.
[6] www.sustainable.org.
[7] See http://www.economist.com/node/21530130.

FIGURE 2-1 Satellite view of shrimp farm ponds in Ecuador. Darker rectangles represent shrimp farm ponds that have encroached upon natural landscapes, including mangrove ecosystems. According to the UN Food and Agricultural Organization, in 1999 Ecuador was the fourth largest producer of shrimp in the world, due almost exclusively to conversion of wetlands to shrimp farms. (Courtesy of NASA.)

mammals have dangerously high levels of toxic artificial chemicals such as organochlorines in their tissue.[8]

Protecting aquatic wildlife could be aided through sustainable aquaculture. For example, growing herbivorous fish like carp and tilapia puts less strain on resources compared to growing carnivorous species like salmon, which usually must be fed with feed made from wild fish. Shrimp farming puts great stress on coastal ecosystems, since mangrove communities are often cleared to make room for shrimp ponds. Figure 2-1 shows rectangular shrimp ponds on the coast of Ecuador, taken by National Aeronautics and Space Administration (NASA). Many oceanic species have been decimated by industrial-style fishing practices as well as by massive national subsidies for fishing fleets.

Ironically, the destruction of ocean fisheries coincides with an increased demand for fish resulting from its recognition as a health food.

Forests and Wood Products

Trees have economic value as a raw material, yet the environmental services provided by forests far transcend the economic value of trees. In addition, trees are important for urban communities and essential for the moderation of global climate. Mature trees maintain desirable microclimates and shelter wildlife. Tropical rainforests actually generate their own precipitation. Figures 2-2a and 2b show Africa's Mt. Kilimanjaro, one of the planet's most awesome volcanoes. The "snows of Kilimanjaro" immortalized by Ernest Hemingway are rapidly vanishing. Figure 2-2b shows the volcano in February of 2000, with snows nearly gone. In addition to global climate change, one of the reasons snowfields are disappearing from Mt. Kilimanjaro is the destruction of forests around and on

[8] See for example http://www.ukmarinesac.org.uk/activities/water-quality/wq8_42.htm.

A

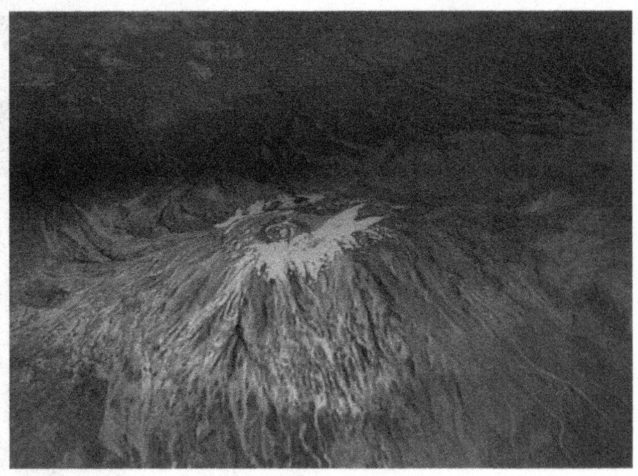

B

FIGURE 2-2 Mt. Kilimanjaro, whose snowcap is suffering from the dual insults of global warming and deforestation. A: Mt. Kilimanjaro's full snowcap in the late 20th century. (Courtesy of Anup Shah/naturepl.com.) B: A view of the volcano in February of 2000 showing the loss of snow. (Courtesy of NASA.)

the giant volcano—forests that transpire sufficient moisture to fuel precipitation, which sustained the snowfields.

In the 13 states of the U.S. Forest Service's Southern Region, forests cover 214 million acres, which amounts to 29% of total forest cover in the United States. Yet, virtually none of the forest in the Region remains pristine, since 99% of southern forests have been cut in the last 400 years. The majority of southern forests are *commercial* forests, typically meaning rows of loblolly pines or other fast-growing trees scattered among recently harvested clearcuts. Known as *tree plantations,* these areas may superficially resemble forests, but they are *monocultures* (composed of a single kind of tree), their biodiversity is not as high as that of undisturbed, natural forests, and they do not provide the same levels of ecosystem services (see below) that natural forests provide and on which humans depend.

Forests protect water supplies and provide habitats, enhancing species diversity. They store carbon, mediating climate change. Healthy forests are essential to sustainable societies. In North America, this means forests with diversity levels approaching those encountered when Europeans first colonized the hemisphere.

Water Supplies

High-quality water supplies are essential both for human use and to maintain the health of local ecosystems. Protection of global water supplies and aquatic ecosystems in the face of growing human populations will be one of humanity's greatest challenges. Reducing waste of water in irrigated agriculture is one of the easiest ways to increase water supplies. Subsidies for irrigation, however, often encourage waste over conservation.

In the United States, Western water laws and doctrines often require those with water rights to use them or lose them. However, change is possible and is indeed underway. In California, agreement has tentatively been reached between irrigators in the Imperial Valley and other stakeholders to share water from the Colorado River and to ultimately reduce water use. However, Mexican farmers, who depend on irrigation water leaking from unlined canals across the border in the Imperial Valley, are suing to prevent sealing of irrigation canals.

Energy

Sustainable societies cannot be built on nonrenewable energy resources. Humans use almost unimaginable amounts of energy and generate vast amounts of pollution. Reducing pollution from fossil fuels requires, at the very least, that laws be strictly enforced. For example, according to the Vermont Journal of Environmental Law, "Chinese environmental laws and regulations are abundant, but suffer from a lack of proper adherence and enforcement," resulting in the needless production of pollutants like SOx and NOx (oxides of sulfur and nitrogen), particulates, and heavy metals like mercury and uranium.[9]

Pollution imposes significant, measurable costs on human and ecosystem health. Energy conservation and the use of renewable fuels provide cost-effective and sustainable alternatives that generate little air and water pollution. Subsidizing the production of coal and oil-based nonrenewable energy makes little sense in a world threatened with rapid climate change and accelerating species loss, in part due to the burning of fossil fuels.

Here, too, change is coming. Wind energy is the fastest growing energy source in Europe and North America, supported by government subsidies which partly offset subsidies for fossil fuels. Energy companies like NextEra Energy and General Electric are major producers and developers of renewable solar and wind energy. And although biofuels cannot yet replace fossil fuels in transportation, producing transport fuels from agricultural waste, and even waste cooking oil, can make a dent in expensive imports and reduce air pollution at the same time.

Biodiversity

Habitat loss is the greatest threat to biodiversity on the planet.[10] Increased food production, including meat, for a richer, more populous Earth is a major cause of habitat loss. Over half of global forests have already disappeared, and they are being removed at a rate, 17 million hectares a year, ten times greater than maximum restoration rates.

Our very survival as a species could ultimately rely on maintaining the integrity of ecosystems we barely understand. An ecosystem is a geographic area including all the living organisms (people, plants, animals, and microbes); their physical surroundings (such as soil, water, and air); and the natural cycles that sustain them (such as the hydrologic cycle). All of these elements are interconnected. Altering any one component affects the others in that ecosystem. Ecosystems can be small, like a single stand of trees, or large, like an entire watershed.

Biodiversity is particularly critical for sustainability because of the specialized and often little understood roles each species plays in maintaining the dynamic state of ecological balance. Moreover, surprisingly little is known about key ecosystems like soils and the deep ocean.

Esthetics and ethics must also play a part since humans can survive, after a fashion, on an Earth with drastically reduced species diversity. The question then becomes, do we wish to make a decision for future generations to eradicate species and ecosystems, without the input of our descendants? That our ancestors did so in ignorance is no excuse for our perpetuating such behavior.

Climate Change

While climate change is a well-documented fact of planetary history—the Earth has gone through several megacycles (100,000,000- to 1-billion-year cycles) of "greenhouse" and "icehouse" conditions—it is the speed with which human-induced climate change

---

[9] VT Journal of Environmental Law, www.vjel.org/journal/pdf/VJEL10058.pdf.

[10] See for example The International Year of Biodiversity, http://www.cbd.int/2010/biodiversity/.

is occurring that is unprecedented. Too-rapid change overwhelms the ability of natural ecosystems to adapt, which is exacerbated by the fragmentation of ecosystems by human activity. For example, the rapid acidification of the oceans will likely help exterminate coral reefs by mid-century unless checked, with serious implications for the entire ocean ecosystem. The impacts of climate change are imperfectly understood, but on the whole will certainly test the creativity and capabilities of a human species that "subdued" a seemingly limitless Earth. Sustainable societies may well be essential to address the impacts of climate change.

Manufacturing and Industry

The Industrial Revolution generated wealth (for some) beyond humanity's dreams, but also generated waste in unprecedented quantities, far beyond the capacity of natural systems to process. Pollution is one form of waste. Wasteless production must become the norm in human activity. And progress is being made: The EU has set a goal of no more waste going to landfills by 2025.

In nature, waste does not exist. Waste eventually becomes something else's food. In human societies, waste is everywhere. Waste indicates inefficiency. Waste can also harm human health and degrade the environment. Many businesses have found that waste reduction and even elimination can enhance profitability. For example, Waste Management Corporation is using landfill gas (methane) to power a sizable proportion of its 22,000 waste collection vehicles,[11] and the Clorox Corp. is phasing out the use of dangerous chemicals in chlorine production. Much progress has been made in this arena—humans have agreed to phase out or eliminate the most harmful kinds of Persistent Organic Pollutants (POPs; see Chapter 12). Many destructive chlorofluorocarbons (CFCs) are being phased out globally under the Montreal Protocol, even though the United States has relentlessly sought exemptions for agricultural users of methyl bromide. However, the growth of human populations and the universal association between increasing wealth and increasing waste poses critical problems for a world, five-sixths of whose population is trying to develop along Western-style free-market lines.

Justice and Equity

The pursuit of justice and equal opportunity are key ingredients in a sustainable civilized society. Examples of injustice are lack of adequate housing, health care, lack of access to education, poor sanitation, an inadequate supply of pure water, exposure to environmental toxins, and environmental degradation related to industrial pollution. Rich societies ignore these issues at their peril.

SUSTAINABLE CONSUMPTION: AN OXYMORON?

We are in the twilight of the era of the "myth of unlimited resources."[12] Humans contribute to local and global sustainability by adopting responsible patterns of buying, consumption, and reproduction, thereby consuming minimal energy and fewer resources. For example, conventional methods of construction do not lend themselves to minimizing energy consumption and waste. Construction and Demolition (C&D) Waste is accordingly a major, and often unnecessary, component of Municipal Solid Waste. Responsible consumption is based on education not coercion in a democratic society. Unfortunately, industrial and postindustrial societies are philosophically devoted to ever-increasing consumption, in turn driving ever-increasing production—the "growth" concept. Detoxifying society

[11] See for example http://www.wm.com/sustainability/renewable-energy.jsp.

[12] See for example http://www.jayhanson.us/page130.htm.

from the "unlimited consumption" myth (perhaps by favoring service consumption rather than material consumption) may be one of our greatest challenges. Sustainable societies are probably incompatible with ever-increasing numbers of "self-storage" facilities, for example.

HEALTH AND NUTRITION

Poor individual physical and mental health imposes significant costs upon society, in the form of health-care expenses, crime, and lost productivity, for example. With all the environmental toxins loosed on the planet by human activity, the greatest killers in wealthy societies remain diseases related to smoking, alcohol, drugs, and obesity. Estimates of the total health and productivity costs of cigarette smoking run as high as $193 billion per year. And the Center for Disease Control and Prevention (CDCP) reports that medical spending on obese persons is $1400 a year more than otherwise. Twenty-seven percent of the U.S. population is now obese and the proportion is growing.[13] While cigarette smoking is declining in America, companies aggressively export the habit to developing countries, a practice that is counter to notions of fairness and equity.

SUSTAINABLE POPULATION

Human numbers must eventually become stabilized, since it is physically impossible for growth to continue forever. The only questions are at what level will growth end and whether growth will end as a result of human actions or by natural processes like famine, disease, and war. Aging societies are typical of developed nations (e.g., the United States). Populations dominated by the young are typical of developing ones (Vietnam, for example). Large numbers of young people provide great promise for societies, but also impose great costs, especially in countries like Spain and Egypt in which up to half the young population of working age (roughly from 16 to 40) is unemployed. Stable, aging societies face challenges of paying for retirement benefits, if such benefits are based on government programs like Social Security, which generally pay to retirees far more in benefits than they paid in.

The readers of this book are mainly young, and it is they who will solve, or not solve, these challenges. The next century should prove to be one of the most interesting, and potentially rewarding, centuries in the entire span of human history.

DEVELOPMENT

A central question in *ecological economics* is to what extent *development* (leading, as it is supposed, to higher per capita income), global *trade*, and a country's *environmental quality* are related.

What Is Development?

Here we use development to refer to a complex set of changes which convert the economy of a society based on subsistence agriculture to one in which most of the employed inhabitants work in manufacturing or services. Early stages of development typically depend more on the exploitation of "natural resources," and later stages on "human capital," that is, the creativity of the human mind.

Development has historically involved significant land-use changes, including

- *deforestation* for fuel wood and to provide land for intensive farming
- *urbanization*, and

[13] See for example http://www.cdc.gov/obesity/data/adult.html.

- *large-scale mining* for fossil fuels and metals.

Development has also produced unprecedented quantities of environmental pollution and waste. These include exhaust gases from the burning of fossil fuels and toxic factory emissions polluting waterways. As some consistently point out, such urban pollutants as horse urine and droppings have declined, but pet waste now constitutes a significant source of urban pollution. For example, there are at least 500,000 dogs in New Jersey (and 8.8 million people) according to that state's Department of Health.

Concentrating humans in cities concentrates waste, both human and otherwise, which if not properly treated can severely pollute waterways or the ocean in the case of coastal cities.[14] Urbanization may also facilitate epidemics. The cholera and yellow fever epidemics in nineteenth century England and North America are examples. However, the concentration of waste and pollution in cities can make it easier to deal with.

Development's Impact on the Environment

Here are two hypotheses, much simplified, that purport to explain such relationships as may exist between development and the environment.

1. **Development harms the environment.** Many environmentalists point out that development leads to harmful land-use practices, injurious levels of air emissions, subsidies encouraging fossil-fuel use, and water pollution, among other things. Moreover, they cite high levels of population growth in many developing countries as exacerbating environmental decline, leading to misery, child prostitution and the like, and encouraging large-scale emigration.
2. **Development eventually improves the environment.** Many economists and some environmentalists, while acknowledging harmful levels of environmental pollution in countries in early stages of development, cite considerable empirical evidence that (1) population growth rates decline as development proceeds and (2) rates of some forms of environmental pollution decline as per capita income increases, a supposed corollary of development as we noted above. Newer forms of technology tend to be less polluting than older forms, but also tend to require high capital expenditures.

Countries with higher per capita incomes tend to have cleaner environments along with increased consumption of goods and services. Poverty and high rates of population growth are major causes of environmental degradation, such as deforestation. As nations become richer and their middle classes expand, so do demands for tougher environmental standards and regulations. Indeed, the "green" movement and "green consumerism" in the developed world evolved with the growth of the middle class after World War II.

ECONOMIC GROWTH AND THE ENVIRONMENT—KUZNETS CURVES

The environmental Kuznets curve, named for Nobel laureate economist Simon S. Kuznets, plots the *relationship between environmental quality factors* and *per capita income*.[15] The relationships that have been plotted include income with the following: sulfur dioxide emissions, suspended particulate matter, carbon monoxide, nitrogen oxides, and airborne lead. In addition, researchers have developed curves by plotting the following environmental parameters against per capita income: access to safe water, presence of urban sanitation

---

14 www.state.nj.us/dep/watershedmgt/pet_waste_fredk.htm.

15 For example see *New York Times* http://tierneylab.blogs.nytimes.com/2009/04/20/the-richer-is-greener-curve/.

(that is, whether or not people are connected to sewage treatment plants), annual deforestation rates, total deforestation, dissolved oxygen in rivers, fecal coliform (a measure of the presence of the toxic bacterium *E. coli*) in rivers, municipal solid waste per capita, and carbon emission per capita.

Shapes of Kuznets Curves

Environmental Kuznets curves (EKC) generally exhibit one of three shapes. One shape results when an environmental benefit improves continually with increasing per capita income. A second shape of Kuznets curves shows a continuous increase in an environmental problem such as municipal solid waste (MSW) with rising incomes. The Kuznets curve that has received the most attention, and has stimulated the most discussion, has an inverted "U" shape.[16] It has been used to suggest the path air quality will follow as economic development, and presumably per capita income, increases. Figure 2-3, depicting the relationship between per capita income and sulfur dioxide level, illustrates one such curve. Similar U-shaped curves have been reported for particulates.

Question 2-2: Interpret the relationship between income and SOx emissions.

Economists differ on income levels at which the "U" shape in the graphs begin. They varied between \$3,000 and \$8,700 for sulfur dioxide and ranged up to \$10,300 for suspended particulate matter.

Question 2-3: Is it reasonable to conclude that some environmental impacts of economic development are not serious because they will decline over time? Why or why not?

Most types of environmental degradation can be offset at a cost. Scrubbers on power plants can remove up to 90 percent of SOx, for example, and increased per capita income gives nations the wealth with which to afford the expense. Carbon emissions per capita show a pattern similar to municipal solid waste; that is, carbon emissions increase with per capita *gross domestic product* (GDP), the total value of goods and services per individual.

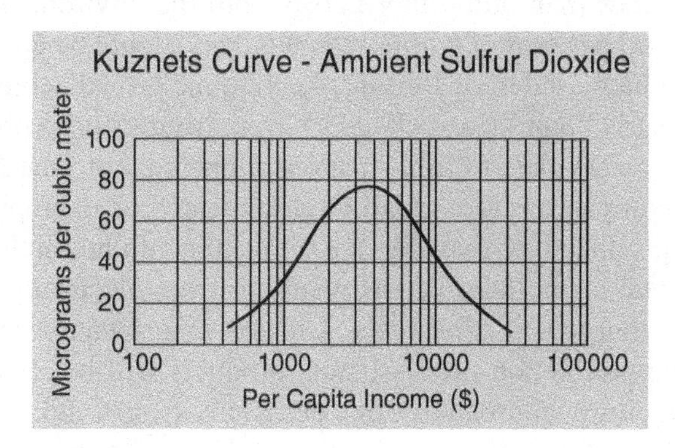

FIGURE 2-3 Kuznets curve showing ambient sulfur dioxide concentration and per capita income.

---

[16] Gluskoter, H. 1997. Some environmental effects of increased energy utilization in the twenty-first century. Proceedings of the 17th World Mining Congress.

Criticisms of Kuznets Curves

Some economists point out that the data are gathered by country, and not globally, and do not take into consideration international trade and the likelihood that wealthier countries are exporting some of their environmental problems to less developed countries. Nor can values for individual countries be extrapolated to the Earth as a whole. However, even when an Environmental Kuznets Curve U-shape relationship is accepted, the *turning point* on the curve, that is, when environmental degradation starts to decline with increasing per capita income, is often found to be very high relative to the per capita GDP of most countries.

In the case of tropical deforestation, researchers found that per capita income levels of most countries in Latin America and Africa were well below the estimated turning point peaks, implying to some that deforestation would eliminate all old-growth forests before those societies could afford to cease the practice.

Such results suggest that the majority of countries have not yet reached levels of per capita income for which environmental improvement is likely to spontaneously occur, unless this improvement is simply dictated by central governments. Worsening global environmental degradation could occur even as the global economy expands and populations grow, and even as some countries make progress cleaning up specific aspects of their environment.

Global Trade and Environmental Quality

The value of international trade in 2010 approached $15.9 trillion, according to the World Trade Organization.[17] The effect of global trade on environmental quality is controversial. Moving oil by tanker can lead to oil spills. And countries can improve their own environment by "exporting" polluting industries and waste to other nations. For example, during the 1970s and 1980s many metal smelters closed in North America, due to the unwillingness of operators to invest in antipollution technology as required by law. Many of these operations simply relocated outside North America to developing countries, who were willing to tolerate the resulting deterioration of their environment. Similarly, the United States and several western European countries have exported toxic wastes like PCBs to countries like Nigeria, where "disposal" costs were a fraction of those in the home country.

Another example of the adverse impacts of global trade is the introduction of *invasive species* into new environments. This can occur either "accidentally" as in the ballast water of cargo ships, or on purpose as investments, or for some presumed benefit.

These and other examples too numerous to cite here represent vast subsidies to world trade—in other words, were the traders required to pay all environmental costs associated with their activities rather than dump those costs onto the environment of the receiving country, the volumes and patterns of world trade would no doubt be considerably different.

However, without increased trade leading to economic growth, countries may not have the financial resources to comprehensively address environmental problems.

At the same time, unregulated "free" markets alone cannot guarantee high levels of economic growth accompanied by a clean environment. For example, costs associated with water pollution produced by manufacturers in China are not included in the price of exported goods. Or total air pollution costs resulting from electricity production are not included in the price charged to consumers. Unless some means can link the pollution abatement costs to the production process, such products will be overproduced and will "out-compete" less-polluting alternatives, like electricity from wind.

---

[17] www.wto.org: see for example www.wto.org/english/news_e/sppl_e/sppl236_e.htm.

Addressing Environmental Impact of Trade and Development

In the case of pollution that crosses national boundaries, addressing the problem usually requires international agreements, and we give two examples below. However, sanctions and punitive tariffs, two oft-proposed remedies, may be counterproductive. As researchers at the Progressive Policy Institute note, "imposing trade sanctions on goods from poor countries with lax environmental standards simply lowers their economic growth and does nothing to counter the poverty that may well be contributing to the environmental problem."[18]Alternatively, international standard-setting organizations could develop voluntary guidelines for labeling, similar to those used to indicate sustainable forestry practices or products of organic agriculture.

Some attempts have been made to address the most egregious problems associated with underregulated trade. The Basel Convention of 1988 began to set controls on the export of toxic wastes from OECD (Organisation for Economic Co-operation and Development) countries to less-developed countries for disposal, for example.

To eliminate the adverse environmental impacts of global trade, many environmental economists, geologists, and ecologists recommend a number of international or multilateral agreements. They include proposals to

- eliminate invasive species from the ballast water of vessels involved in commercial activity in international waterways
- eliminate or tax transboundary air and water pollution
- require "double-hulled" tankers for the shipment of petroleum—already required for shipment of oil in U.S. territorial waters
- promote negotiations to ensure that workers engaged in export industries receive benefits and wages that are at parity with those in the countries to which the goods are shipped
- promote international agreements to regulate such degrading activities as clearcutting and metal smelting so that industries cannot be subsidized by lax environmental standards for moving their polluting activities from one country to another, and
- promote treaties to eliminate, price, or restrict the use of toxic materials, or harmful practices, in agriculture.

The World Trade Organization (WTO)

The founding charter of the WTO formally addresses the relationship of trade to the environment.[19] It states that signatories to the WTO should "allow for the optimal use of the world's resources in accordance with the objective of sustainable development, seeking to both protect and preserve the environment." WTO rules also allow countries to impose trade regulations "necessary to protect human, animal, or plant life or health" or "relating to the conservation of natural resources." However, measures taken to protect the environment must not discriminate. A country may not be lenient with its domestic producers and at the same time be strict with foreign producers. Nor can member nations discriminate among different trading partners.

The WTO has been controversial since its founding. We will use one example to illustrate the controversy, the issue of sea turtles.[20] Five Asian nations challenged a U.S. law designed to protect sea turtles from certain harmful fishing practices. The law banned the importation of shrimp from countries that did not require the use of turtle-excluder devices (TEDs) by their fishing industry. The WTO dispute panel ruled in favor of the Asian countries, not because it disapproved of United States' attempts to protect sea turtles, but

---

[18] For example, see http://environmentaleconomics.wordpress.com/.

[19] See www.wto.org/english/docs_e/docs_e.htm.

[20] See for example International Environmental Law Project http://law.lclark.edu.

because the panel found the United States had discriminated among members of the WTO, granting preferential treatment to Latin American and Caribbean nations, but not to Asian countries. This decision infuriated environmentalists even though it had arguably nothing to do with the desirability of saving sea turtles.

The WTO cannot dictate national government policy. Sovereign nations choose to become members of the WTO and to play by its rules. So far, more than 140 countries have joined, and others have applied for membership.

International Agreements to Protect the Environment

Two examples of relatively successful international agreements to protect the global environment are the Basel Accords on Toxic Waste and the Montreal Protocol.

The Basel Accords on Toxic Waste. The Basel Accords on Toxic Waste[21] became effective in 1992. It identified nearly four dozen specific categories of waste to be deemed toxic, and banned export of any of these materials from any signatory nation to any signatory nation that indicated that it did not want to receive such waste. It contained articles defining illegal activity in waste shipment and required all signatory states to eliminate such activity. It stipulated that such waste should be minimized, and disposal or treatment should occur as close to the source as possible.

As of 2010, most nations, including the EU, had ratified the Convention. Three signatories to the Convention had not yet ratified it: Afghanistan, Haiti, and the United States.

The Montreal Protocol. Severe depletion in stratospheric ozone has been measured for years, especially in the Southern Hemisphere. The Montreal Protocol on Substances that Deplete the Ozone Layer was adopted in 1987 as an international treaty to eliminate the production and consumption of ozone-depleting chemicals (ODCs). Four agencies were tasked with implementing the Protocol: the World Bank and three United Nations' agencies—the UN Environment Programme, the UN Development Programme, and the UN Industrial Development Programme.

The Montreal Protocol stipulates that the production and consumption of compounds that deplete ozone in the stratosphere—CFCs, halons, carbon tetrachloride, among others—were to be phased out by 2000, or 2005 in the case of methyl chloroform and methyl bromate. By 2009, the Protocol had achieved universal acceptance, and had achieved a 98% of ozone-depleting chemicals. By mid-2010, developing countries had phased out more than 270,000 tonnes of ozone depleting chemicals.[22]

Conclusion

World trade has enormous potential to foster the objectives of development, but can also be the source of massive environmental degradation, without multinational and international agreements to "level the playing field." The Basel Convention and Montreal Protocol are two examples of the types of international agreements that could serve as templates for new agreements to address the negative impacts of underregulated international trade.

The debate about the impacts of development continues. Kuznets curves are one, albeit controversial, means of analysis by which data on the consequences of development may be evaluated.

---

[21] www.basel.int.

[22] See for example U.S. EPA http://www.epa.gov/ozone/intpol/.

CHAPTER 3

Greenhouse Gases and Climate Change: Part One

KEY QUESTIONS

- What is the composition of the Earth's atmosphere?
- How has the atmosphere changed over planetary history?
- What processes influence and alter the atmosphere's composition?
- What are greenhouse gases, and how do they affect global temperatures?
- How is global sustainability affected by a changing climate on the scale forecasted by atmospheric scientists?

FACTS ABOUT THE ATMOSPHERE

All life on Earth depends on our atmosphere. Despite this, humans are altering the atmosphere's composition with growing understanding, at least on the part of scientists, of medium- to long-term adverse impacts. Scientific evidence has confirmed that emissions from the burning of fossil fuels, from industrial sources such as cement manufacture, and from deforestation have changed and continue to alter the makeup of our atmosphere. In addition, trace gases such as methane and chlorofluorocarbons (CFCs and HCFCs) are having an impact on the atmosphere wholly out of proportion to their concentration. To analyze the effects of human activity on the atmosphere, there are some facts you need to know.

Earth's atmosphere is a thin shell. Half the mass of the atmosphere is within 5 km (3.1 mi) of the surface, and 90 percent is within 20 km (12.4 mi). Twenty km is not very far—about a 10-minute drive on most interstate highways. By comparison, the Earth's radius is 6,378 km (3963 mi).

Question 3-1: What percent of the Earth's radius is 20 kilometers?

The major gases that make up the Earth's atmosphere are shown in Table 3-1. The proportions of the gases (ignoring water vapor, which is variable) in the atmosphere remain virtually constant to a height of about 25 km.

TABLE 3-1 ■ Major Gas Composition of the Atmosphere

Gas	Volume (%)	Molecular Weight (approximate)	Weight (%)
Nitrogen	78.000	28	75.00
Oxygen	21.000	32	23.00
Argon	00.930	40	1.30
Carbon dioxide	00.035	44	0.05
Average atmosphere	100.000	29	100.00

*Percentages do not add up to 100 due to rounding, approximation, and ignoring water vapor and all minor and trace gases.
Weight (%) is found by multiplying the volume (%) by the molecular weight of that gas and dividing by the average molecular weight of the atmosphere.

WEIGHING THE ATMOSPHERE

At sea level, the weight of a 1 cm^2 column of air is approximately 1 kilogram. This value, 1 kg/cm^2, is also known as the atmospheric *pressure*. For this analysis, we will assume a featureless Earth without continents. This will simplify, but not significantly alter, your calculations.

Question 3-2: How much does the atmosphere weigh over each square meter of surface? Express this in tonnes (units of 1000 kg). Recall the conversion factors: 10^4 cm^4 = 1 m^4; 1 tonne = 1000 kg = 10^6 g.

Question 3-3: How much does the atmosphere weigh over each square kilometer (km^2)? Recall that 10^6 m^2 = 1 km^2.

Question 3-4: The area of the Earth's surface is approximately 516×10^6 km^2. What is the weight of the entire atmosphere in tonnes? Express your answer using scientific notation, in units of 10^{12} tonnes.

It is worth pointing out that this is by no means an incomprehensibly large number. The mass of the Earth is 6×10^{27} g. How does the atmosphere's mass compare with that of the Earth?

EARLY PLANETARY ATMOSPHERES

Our current atmosphere is believed to be the Earth's third over its cosmic history. The earliest was comprised of light gases, primarily hydrogen (H_2) and helium (He_2). This first atmosphere was blown away as the Sun reached critical mass and began internal fusion, and by heat generated by the early molten earth, between 4 and 5 billion years ago. The second atmosphere was formed around 4 billion years ago when gases escaped from the Earth as its crust solidified. Bombardment by comets may have also contributed gases to the Earth's second atmosphere. This second atmosphere lacked free oxygen, a gas necessary to support most of the Earth's present life forms.

THE DEVELOPMENT OF THE CURRENT ATMOSPHERE

The Earth's current atmosphere was produced partly by the metabolism of living organisms. Cyanobacteria, primitive photosynthesizing organisms, had appeared by 3.8 billion years ago and began to produce oxygen as a byproduct. Over the next 2 billion years, atmospheric oxygen (O_2) concentrations rose and CO_2 concentrations fell as a direct result of photosynthesis, as well as the sequestration (storage or "fixing") of carbon in *carbonate* rock (limestone and dolostone: $CaCO_3$ and $CaMg[CO_3]_2$). This increase in atmospheric oxygen set the stage for two major evolutionary events on the planet: the evolution of aerobic (oxygen-using) life forms and the formation of the *ozone layer.*

Increasing oxygen in the hydrosphere and atmosphere was toxic to sensitive *anaerobic* microorganisms and prompted the evolution of new microorganisms capable of more efficient respiration, using oxygen to liberate energy from organic compounds. Today, large anaerobic microbial communities are restricted to Mid-Ocean Ridge hydrothermal systems and other marginal environments. The development of aerobic metabolism also permitted the evolution of multicellular organisms, which required more energy to support their increased biomass. All existing multicellular life forms employ aerobic metabolism.

Increasing atmospheric oxygen reacted with high-energy solar radiation and eventually triggered the formation of a layer of ozone (O_3) in the upper atmosphere. Although a pollutant at ground level, the ozone layer filters out harmful high-energy ultraviolet radiation, which can cause skin cancer in humans.

Indeed, before the formation of the ozone layer, the Earth's surface was exposed to extremely high intensities of ultraviolet radiation. The surface ocean waters filtered out some of this radiation and thus provided some protection to organisms, but it is likely that ocean primary production (that is, production of high-energy compounds from photosynthesis) was still limited by the high ultraviolet radiation intensities. The terrestrial surface fared worse and was possibly even sterilized by this radiation: The first widely accepted evidence for land plant communities, for example, appears in rocks that are only about 400 million years old.

FUNCTIONS OF THE ATMOSPHERE

In addition to providing the oxygen needed by most of Earth's life forms, the atmosphere provides thermal insulation, preventing extreme changes in temperature over the daily light-dark cycle. Unequal heating of the Earth's atmosphere and terrestrial surface create long-term *climate* and short-term *weather* patterns. The winds that result from these heating differences and resultant pressure differences help drive ocean currents. The atmosphere also transfers vast quantities of heat from equatorial to polar latitudes.

THE ATMOSPHERE'S CURRENT COMPOSITION

As Table 3-1 shows, the present-day atmosphere is composed primarily of nitrogen (N_2) gas (78.08% by volume), oxygen (O_2; 20.94%), and argon (Ar; 0.93%). It also contains water vapor (H_2O), carbon dioxide (CO_2), neon (Ne_2), helium (He_2), methane (CH_4), oxides of sulfur (SOx) and nitrogen (NOx), and ozone (O_3). The major controls on the composition of the atmosphere, and the cycling of these compounds in and out of the atmosphere, are interactions with the Earth's biosphere (living matter), hydrosphere, and lithosphere (rock and geological processes such as volcanism).

Presently, O_2 levels seem to be stable, but CO_2 levels are not and display seasonal and longer-term trends. Seasonal changes in CO_2 are related to seasonally fluctuating primary production (i.e., plant growth) due to changing light durations. Longer-term (decade-to-century) increases in CO_2 are due to a variety of anthropogenic (human-caused) inputs, including changes in land use that reduce the ability of terrestrial biota to absorb CO_2. The present concentration of CO_2 is higher than it has been for at least 400,000 years, according to the Intergovernmental Panel on Climate Change (IPCC).

Because the amount of CO_2 in the atmosphere is very small, the concentration is easily changed by the addition of CO_2 from various sources.

CHANGES CAUSED BY HUMANS

After decades of research, scientists have concluded that humans have altered the composition of the Earth's atmosphere. Before the Industrial Revolution (ca. 1750), clear-cutting of forests in Europe, China, and the Middle East, and later in North America, set the stage for modifying the atmosphere's composition. Cutting and burning forests liberates CO_2 in two ways. Carbon from the vegetation is converted to CO_2 and soils, devoid of their tree cover, emit CO_2 at greater rates than before.

Since the beginning of the Industrial Revolution, atmospheric concentrations of greenhouse gases have increased. Carbon dioxide has risen by one third; methane has more than doubled; and nitrous oxide concentrations have risen by about 15 percent (Table 3-2).

The Greenhouse Effect

Here's how the *greenhouse effect* works. Greenhouse gases allow short wavelength radiation (mainly visible light and ultraviolet) from the Sun to pass through the atmosphere, but they absorb the longer wavelength radiation (mainly infrared) that is emitted by the Earth's surface, thereby heating the atmosphere. Over the past two centuries, (1) the global human

TABLE 3-2 ■ Changes in the Global Concentration of Greenhouse Gases Since the Preindustrial Period[1]

	CO_2	CH_4	N_2O
Preindustrial concentration	280 ppmv	700 ppbv	275 ppbv
Concentration in 2005	380 ppmv	1720 ppbv	312 ppbv
Concentration in 2011	391 ppmv	1810 ppbv	332 ppbv
Atmospheric lifetime (years)	50–200	12	120

ppmv = per million by volume; ppbv = per billion by volume

[1]Intergovernmental Panel on Climate Change (IPCC). 1995: The science of climate change: Contribution of Working Group I to the Second Assessment Report to IPCC on Climate Change. J.T. Houghton, L.G. Meira Filho, B.A. Callander, N. Harris, A. Kattenberg & K. Maskell, eds. (Cambridge University Press: New York). Courtesy of J. T. Houghton, L. G. Meira Filho, B. A. Callander, and N. Harris/IPCC Secretariat.

Reference: http://cdiac.ornl.gov/pns/current_ghg.html, Oak Ridge National Lab.

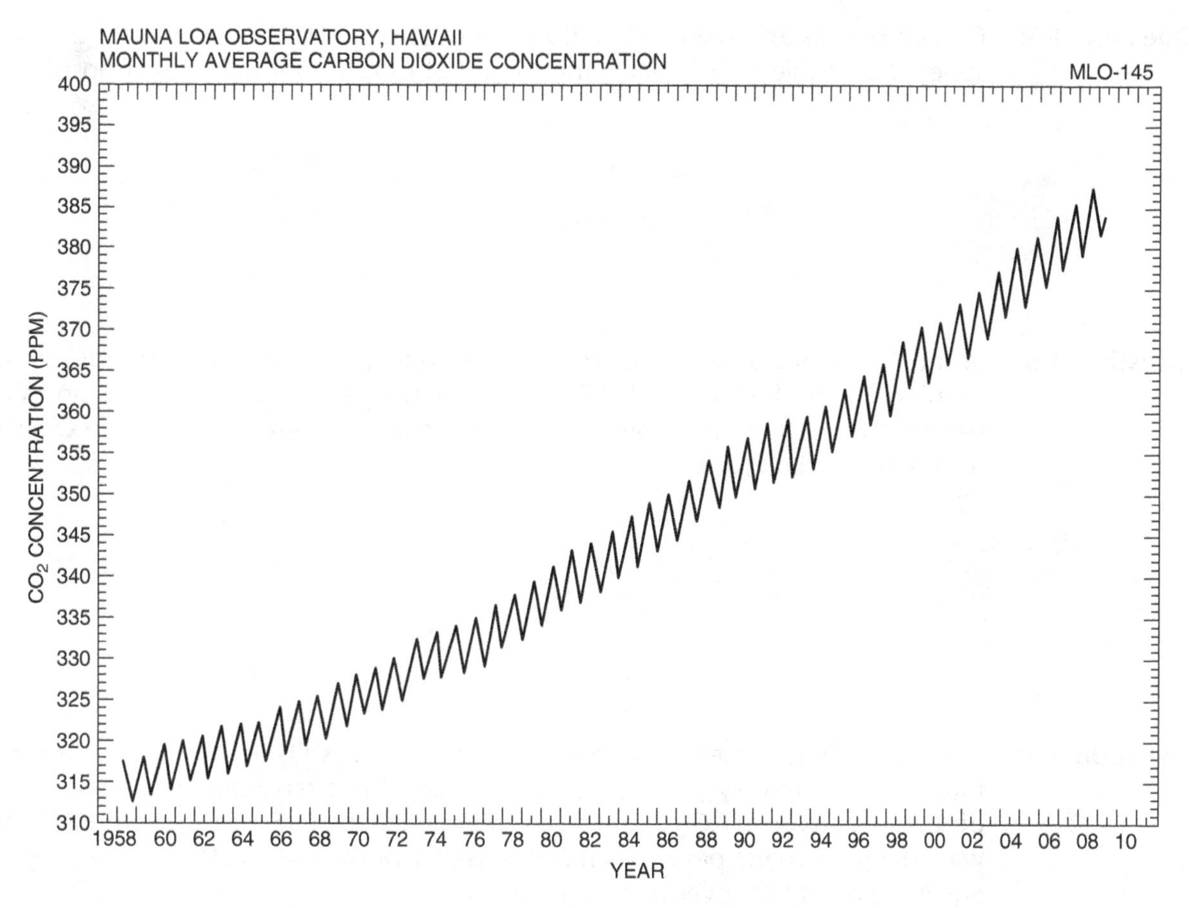

FIGURE 3-1 Atmospheric CO_2 concentrations measured at the Mauna Loa Observatory from 1958 to 2010. (Source: C.D. Keeling and T.P. Whorf. Atmospheric carbon dioxide record from Mauna Loa, http://cdiac.esd.ornl.gov/trends/co2/sio-mlo.htm and http://cdiac.ornl.gov/pns/current_ghg.html)

population has grown tenfold; (2) the demand for energy to support industrial development, heat homes, cook food, watch television, and surf the Internet has vastly grown; and (3) the increased use of automobiles has resulted in the burning of great stores of fossil fuel. Figure 3-1 shows what is known as *the Keeling Curve*, the change in atmospheric CO_2 concentration over time.

Fossil fuels, including coal, oil, and natural gas, were formed by the burial and slow anaerobic decomposition of ancient plant and phytoplankton deposits. While these deposits took tens of millions of years to form, we are burning them at a vastly more rapid rate. A key by-product of fossil fuel consumption is CO_2. Since the Industrial Revolution, we have added CO_2 to the atmosphere far more rapidly than it can be absorbed by its variety of sinks, or "storehouses." This has led to a steady increase in CO_2 concentration that will result in at least a doubling of pre-1860 atmospheric CO_2 content by the year 2150, if present trends continue.

Methane is another greenhouse gas whose concentration has increased because of human activities. Methane is emitted by burping cows, flooded farmlands (i.e., rice paddies), coal seams, leaking gas pipelines, and municipal landfills.

CARBON DIOXIDE IN THE ATMOSPHERE

Previously, you calculated the weight of the entire atmosphere. Now let's figure out how much of this weight is CO_2.

Question 3-5: CO_2 comprises approximately 0.05 percent (1/20 of 1%) by weight of the atmosphere (see Table 3-1). What is the weight of CO_2 in tonnes in the atmosphere?

Question 3-6: World CO_2 emission into the atmosphere alone from deforestation, industry, and burning of fossil fuels exceeds 18.4 billion tonnes per year and is increasing.[1] What percentage of the total amount of CO_2 in the atmosphere (calculated in Question 3-5) is 18.4 billion tonnes?

Question 3-7: Thus, according to these calculations, we are adding CO_2 to the atmosphere at the rate of approximately _____ percent per year. At this rate, can we account for the observed increase in CO_2 in the atmosphere since the 1950s? CO_2 content in 1959 was 316 ppm (parts per million) and was 391 ppm in 2011. What is the percentage increase in CO_2 over this interval?

Question 3-8: Is this percentage increase within the order of magnitude suggested by your calculations? (Order of magnitude means roughly within the same power of 10; e.g., the ratios 3.5 cm/yr and 6.5 cm/yr are within the same order of magnitude, whereas 3.5 cm/yr and 26.8 cm/yr are not.)

Question 3-9: Recall the introduction to this chapter: Do you think all of the CO_2 produced by human activities stays in the atmosphere? Identify a major "sink," or storehouse for this anthropogenic (human-generated) CO_2.

---

[1] http://co2now.org/ and www.iea.org.

In the next chapter, we will evaluate the nature of the climate change risk we face and what, if anything, we can do to mitigate it. For now, think about the importance of the rate at which climate changes.

Question 3-10: Summarize the major points of this chapter.

Question 3-11: Discuss the issue of greenhouse gases and climate change from the standpoint of sustainability.

FOR FURTHER THOUGHT

Question 3-12: Research the carbon cycle. Discuss the importance of limestone and fossil fuels as carbon storehouses (sinks).

Question 3-13: Many countries with sizable populations living in coastal areas, as well as some small countries with their entire land area near or at sea level, are, or should be, acutely concerned about the impact of sea-level rise associated with human alteration of the atmosphere. The Maldives are one such country, and Bangladesh was discussed. But the countries with most to lose are probably China, Japan, and perhaps Vietnam, with hundreds of millions of people and trillions of dollars of real estate at risk by 2080. How are China, Japan, and Vietnam dealing with the risks of climate change?

Question 3-14: Research the impact of "Superstorm" Sandy (October 2012) upon the Northeastern United States. Why was this storm so dangerous? Do climate scientists think that the storm could have been caused by climate change? Explain your answer.

Chapter 4

GREENHOUSE GASES AND CLIMATE CHANGE: PART TWO

Key Questions

- How fast is the Earth's climate changing?
- What impacts of global climate change will result from greenhouse gas increases?
- How does the Earth's atmosphere interact with the ocean?
- How can climate change affect ocean circulation?
- Can human-induced global climate change be reversed?

During the summer of 2012, thousands of temperature records were broken across the United States from the Rocky Mountains to the Eastern Seaboard. An expected bumper corn crop was decimated by extreme heat and drought. The West Coast temperatures were normal. Are these harbingers of a new normal?

Question 4-1: Access http://www.globalchange.gov/images/cir/hi-res/11-southeast-pg-112_top.png, an official government forecast of the increase in very hot days in the American Southeast due to human-induced climate change. How many more days above 90°F is eastern Virginia likely to experience?

Introduction: The 3rd IPCC Assessment

At a meeting in Shanghai in January 2001, the Intergovernmental Panel on Climate Change (IPCC) issued its most comprehensive report to date on the environmental implications of climate change. Over 150 delegates from nearly 100 governments met to consider the Third Assessment Report of the IPCC "Climate Change 2001: The Scientific Basis."[1] The full report, which runs to over 1,000 pages, is the work of 123 lead authors, who used the contributions of more than 500 scientists. The report went through extensive review by experts and governments. In line with IPCC Principles and Procedures, after line-by-line consideration, the governments unanimously approved the Summary for Policymakers of the report and accepted the full report.

[1] Available at www.grida.no/climate/ipcc_tar/.

Here are the major conclusions of this report:[2]

1. Confidence in models of future climate has increased. Climate data for the past 1,000 years, as well as model estimates of natural climate variations, suggest that there is an anthropogenic "signal" in the climate record of the last 35–50 years, meaning this change has certainly been affected by human activities. Analyses of tree rings, corals, ice cores, and historical records for the Northern Hemisphere indicate that the increase in temperature in the twentieth century is likely to have been the largest of any century during the past 1,000 years. It is likely that the 1990s were the warmest decade and 1998 was the warmest year during the past millennium.
2. In the mid- and high-latitudes of the Northern Hemisphere snow cover has likely decreased by about 10 percent since the late 1960s. The annual duration of ice cover shortened by about two weeks over the twentieth century. It is likely that there has been about a 40 percent decline in Arctic sea-ice thickness during late summer to early autumn in recent decades.
3. Since 1750, the atmospheric concentration of carbon dioxide increased from 280 parts per million (ppm) to about 367 ppm in 1998.[3] *The present* CO_2 *concentration has not been exceeded during the past 420,000 years and likely not during the past 20 million years* [our emphasis].
4. The global averaged surface temperature is projected to increase by 1.4–5.8°C from 1990 to 2100. This is higher than the 1995 Assessment Report's[4] projection of 1–3.5°C, largely because future sulfur dioxide emissions (which help to cool the Earth) are now expected to be lower—the result of lower air pollution. (However, during the 2000s particulate emissions from massive Chinese coal plants may have started to cool the Earth by reflecting and diffusing incoming solar radiation—a phenomenon not considered in early climate models.) This future warming is on top of a 0.6°C increase since 1861.
5. Global average water vapor concentration (humidity) and precipitation are projected to increase. More intense precipitation events are likely over many Northern Hemisphere mid- to high-latitude land areas. There is renewed debate on the subject after the 2005–2006 hurricane seasons.
6. Sea level was projected to rise by 0.09 to 0.88 meters from 1990 to 2100.

UPDATE: THE 2007 (4TH) IPCC ASSESSMENT AND BEYOND

The 2007 IPCC Assessment[5] largely confirmed the findings of the 3rd Assessment, with these new findings:

1. The warming of the climate system is now "unequivocal."
2. Of the twelve years between 1995–2006, eleven were among the warmest ever measured (since 1850).
3. Sea level rise since 1993 was estimated at 2.4–3.8 mm/y, with an average rate of 3.1mm/y.

---

[2] Adopted with changes from "New Evidence Confirms Rapid Global Warming, say Scientists." UNEP News Release 01/5. Available at www.unep.org/Documents/Default.asp?DocumentID=296.

[3] Keeling, C.D. & T.P. Whorf. Atmospheric carbon dioxide record from Mauna Loa, http://cdiac.esd.ornl.gov/ trends/co2/sio-mlo.htm. By 2012 CO_2 had reached 391 ppm.

[4] Intergovernmental Panel on Climate Change (IPCC). 1995. Climate Change 1995: The science of climate change: Contribution of Working Group I to the Second Assessment Report of the Intergovernmental Panel on Climate Change. J.T. Houghton, L.G. Meira Filho, B.A. Callander, N. Harris, A. Kattenberg & K. Maskell, eds. (Cambridge University Press: New York).

[5] http://www.ipcc.ch/.

4. Average Northern Hemisphere temperatures during the last half of the 20th century were likely higher than at any time in the past 1300 years. There is "medium confidence" that "other effects" of climate change are emerging, including heat wave mortality in parts of Europe, and the expansion of the range of formerly tropical disease vectors.
5. The West Antarctic Ice Sheet contains enough ice to raise sea level by 5–6 meters. There is a "small chance" that the collapse of this sheet could occur during the next few centuries.

There is a scientific consensus that the principal drivers of current, accelerated climate change are burning of fossil fuels and deforestation. For the latest information, start at the following web sites: U.S. EPA Climate Change (http://www.epa.gov/climatechange/), NASA Global Climate Change (http://climate.nasa.gov/), and the Intergovernmental Panel on Climate Change (www.ipcc.ch/).

Question 4-2: Visit any of the above web sites and summarize current knowledge on the impacts and science of climate change.

IMPACTS OF GLOBAL CLIMATE CHANGE

Examine Figure 4-1 for an overview.

1. *Human health* will be directly affected by increases in the *heat index*, which especially affects the elderly and those with heart and respiratory illnesses. (See above.) Higher temperatures will also increase ozone pollution in the lower atmosphere, a further threat to people with respiratory illnesses.
2. Warming may increase the incidence of some infectious diseases, particularly those that usually appear only in warm areas. Diseases that are spread by "vectors"—mosquitoes and other insects, etc.—such as malaria, dengue fever, yellow fever, and encephalitis, could become more prevalent if warmer temperatures and wetter climates enable those insects to become established farther poleward. Already, some municipalities are planning increased spraying of insecticides to combat tropical vector-borne (mainly by mosquitoes) diseases.
3. Alterations in precipitation patterns are likely, and in fact may already be occurring according to the 4th Assessment. Changes in climate are expected to enhance both evaporation and precipitation in most areas of the United States. The net balance of evaporation and precipitation influences the availability and quality of water resources. In areas expected to become more arid, like California, lower river flows and lower lake levels could impair navigation, reduce hydroelectric power generation, decrease water quality, and reduce the supplies of water available for agricultural, recreational, residential, and industrial uses. In other areas, increased precipitation is expected to be more concentrated in large storms as temperatures rise. This could increase the incidence and severity of flooding.

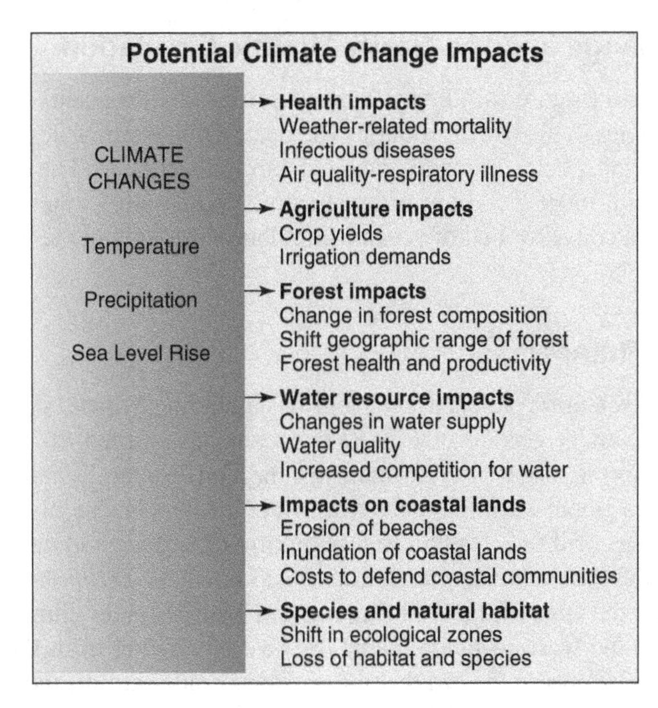

FIGURE 4-1 Impacts of climate change.

4. Climate change could have severe impacts on agriculture (e.g., increasing evaporation from soils, thus drying them out). These impacts could be partially offset, at least in the United States, by longer growing seasons and enhanced crop production from higher atmospheric CO_2.
5. Climatic change will cause a shift in biotic community composition as plant and animal species try to migrate to maintain their preferred habitats. For example, the projected *average* 2+°C (3.6°F) warming likely to occur this next century could shift the ideal range for many North American forest species by about 300 kilometers (200 mi) to the north. Many plant species lack seed dispersion mechanisms that can adjust this rapidly, and these species may become regionally extinct. Coastal wetland plant communities may lose habitat because they may not be able to keep up with the predicted rate of sea-level rise, or the migration paths of these communities may be blocked by human development. Coral reef ecosystems, already under stress from human impact, may be reduced to a fraction of their historical range.

Wildlife species may be severely impacted by climate change[6]. For example, some duck species are dependent upon "prairie potholes" found in the Northern Great Plains. A drier climate would decrease the number and size of ponds in this region, with a commensurate reduction in duck populations. Fish that inhabit inland aquatic environments will be more vulnerable than coastal or marine species. Lake-locked fishes have little recourse in seeking cooler waters. Fish that inhabit north-south rivers may be able to migrate to seek cooler water, but fish in east-west oriented rivers and lakes will not be able to escape warming impacts. The diversity of fishes in U.S. rivers and streams is likely to decline, according to one study[7] in some local systems as much as 75%.

[6] Bellard, C. et al. 2012. Impacts of climate change on the future of biodiversity. *Ecol. Lett.* 15: 365–377.

[7] Xenopoulos, M.A. et al. 2005. Scenarios of freshwater fish extinctions from climate change and water withdrawal. *Global Change Biol.*, 11, 1557–1564.

Climate Change and the North Atlantic Circulation

The Gulf Stream (Figure 4-2) brings vast quantities of warm, salty water from the tropics to higher latitudes in the North Atlantic. The heat in this warm water, coupled with prevailing westerly winds, helps moderate much of Europe's climate. The entire system seems to depend on rapid, intense cooling of hypersaline (extra-salty) water off Greenland, which sets in motion a conveyor belt of oceanic circulation throughout the world's oceans (Figure 4-3).

A Disappearing Gulf Stream?

In a number of scientific papers in such journals including *Nature, Science,* and *GSA Today,* Columbia University geoscientist W.S. Broecker has warned of "an oceanic flip-flop." Global warming, he argues, could interrupt the Gulf Stream's *thermohaline* circulation. Here's how: As global warming melts freshwater glaciers in Greenland, salt concentrations in surface waters will fall, leading to less mixing of surface and deep waters. This would stop or slow the thermohaline circulation (also known as *meridional overturning circulation*), interrupt the flow of the Gulf Stream, and bring a cooler climate to northern Europe as winds from the west no longer carry the heat emanating from the Gulf Stream. More recently, Broecker has suggested that these effects could turn off the deep-ocean conveyor belt completely, possibly triggering intense cooling in Europe. Evidence for such flip-flops has been found in geological records obtained from ice cores and deep-sea sediments. Of particular concern is the fact that these events have occurred over time periods as short as *four years.* Broecker calls the oceans "the Achilles' heel" of the climate system.

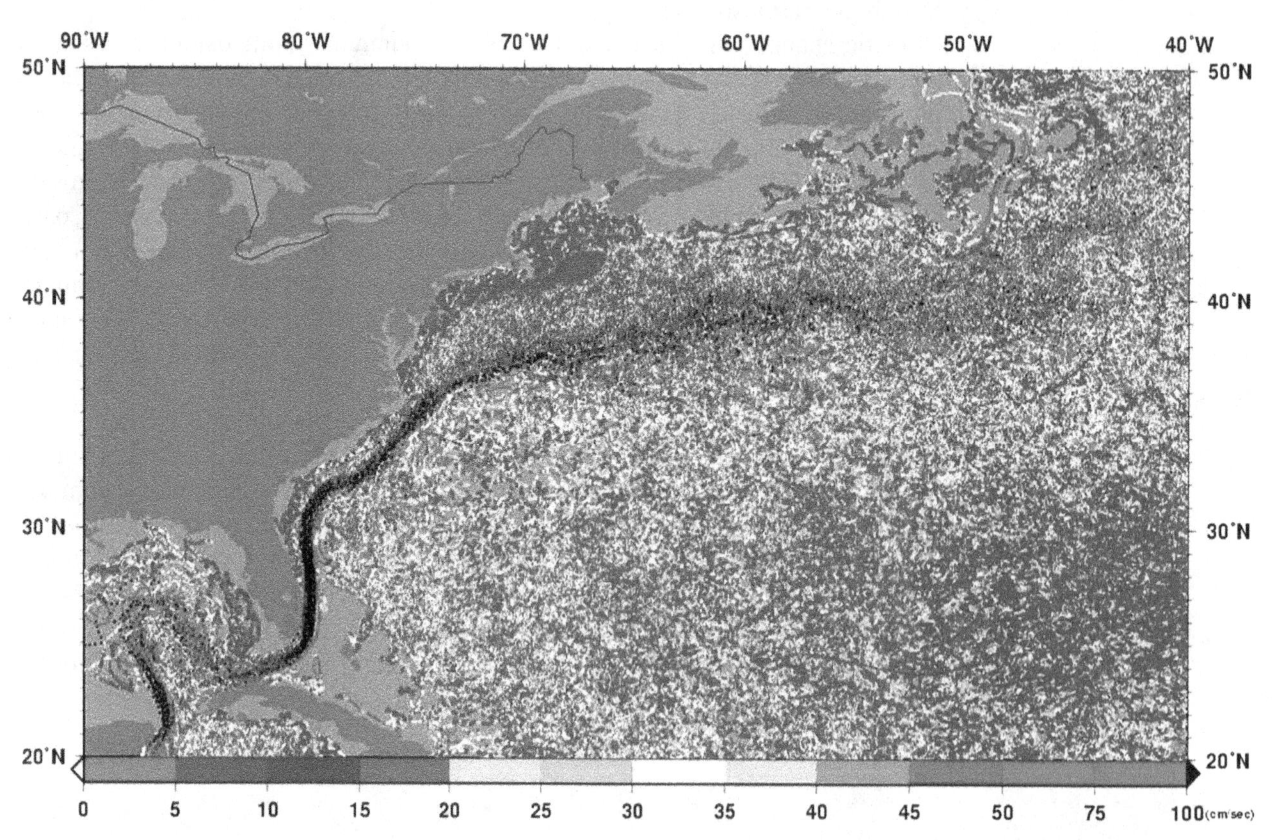

FIGURE 4-2 The Gulf Stream. (Professor Arthur J. Mariano Phd, University of Miami, http://oceancurrents.rsmas.miami.edu. Courtesy of Rosenstiel School of Marine and Atmospheric Science.)

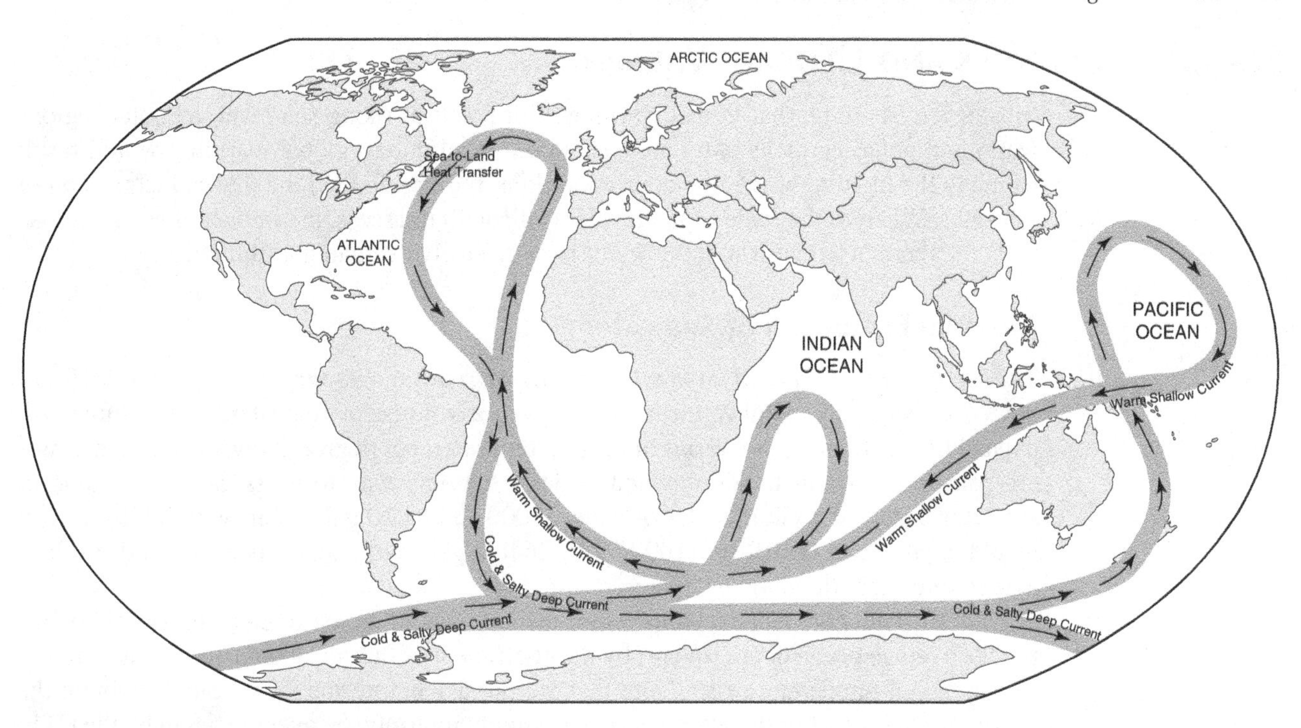

FIGURE 4-3 Conveyor-belt circulation. This circulation is initiated when cool North Atlantic water sinks. It then flows southward and moves into the Indian and Pacific Oceans.

Question 4-3: In a 1999 paper in *GSA Today*, Broecker says, ". . . an extreme scenario is an unlikely one, for models suggest that in order to force a conveyor shutdown, Earth would have to undergo a 4 to 5 degree C warming."[8] Based on this statement and Item 4 (p. 37) of the 2001 IPCC report, comment on the possibility of a conveyor shutdown in the North Atlantic before 2100.

Question 4-4: Sulfur dioxide (SO_2) is a pollutant produced mainly by coal and oil combustion. It is a toxic chemical in high concentrations and is a major contributor to acid precipitation. Pollution from Chinese coal plants puts vast quantities of SO_2 into the upper atmosphere, where it may help to explain why global temperatures have not been as high over the past half-decade as models forecast. (SO_2 in the upper atmosphere disperses incoming sunlight and thus may mediate atmospheric temperature increase.) Would you favor relaxing pollution reduction regulations to help reduce global warming, as some have proposed? Discuss.

[8] Broecker, W.S. 1999. What if the conveyor were to shut down? *GSA TODAY, 9* (1): 2–7.

SEA-LEVEL CHANGES AND GLOBAL WARMING

Most scientists think that global warming will result in rising sea level, as polar regions warm and polar ice melts. Any increase in sea level due to global warming would result from (1) the melting of ice in glaciers and polar regions and (2) the thermal expansion of seawater. Whereas the scale of the former is difficult to assess, the impact of sea-level rise from the thermal expansion of the water is more easily evaluated arithmetically.

Thermal Expansion of Seawater

The *coefficient of thermal expansion (CTE)* of seawater is approximately 0.00019 per degree Celsius. This simply means that, given a volume of seawater, that volume will expand by this fraction as a result of heating the water, per degree. Now since ocean basins are constrained by their bottoms and "sides," the only way to go is up. If a volume of seawater occupied 1 cubic meter of water (1000 L, 264.20 gal), after warming by 1°C, it would expand to 1.00019 m^3 (1000.19 L, 264.25 gal). Note that this also could result in extensive coastal flooding.

To calculate how much a temperature increase would increase sea level, simply multiply the average ocean depth (in cm) by the coefficient of thermal expansion by the number of degrees of temperature rise. Note that we put a 1 before the CTE value to obtain the height of sea level after the warming. If you simply multiply the average depth by the CTE, you will get only the number of centimeters that sea level will rise.

Question 4-5: How much would each 1°C increase in seawater temperature cause sea level to rise? Express your answer in cm (30 cm = 1 ft). Recall the average depth is 3,800 m (3.8 km).

Whereas thermal expansion acts upon water already in the basin, melting ice represents water added to the present ocean volume. The melting of ice that is currently perched upon terrestrial land, as in the ice sheets of Greenland, Iceland, and the Antarctic, has the potential to raise sea level by about 80 meters (260 ft). Ice that is already floating in the ocean water (Arctic ice, Antarctic ice shelves, and icebergs) may melt but will not contribute to sea-level rise. The mass of water contained in these features already displaces approximately its equivalent water volume.

A recent EPA report concluded, "The total cost for a one meter rise would be $270–475 billion (in the U.S.), ignoring future development. We estimate that if no measures are taken to hold back the sea, a one meter rise in sea level would inundate 14,000 square miles, with wet and dry land each accounting for about half the loss. The 1500 square kilometers (600–700 square miles) of densely developed coastal lowlands could be protected for approximately one to two thousand dollars per year for a typical coastal lot. Given high coastal property values, holding back the sea would probably be cost-effective."[9]

[9] www.epa.gov/climatechange/effects/coastal/slrmaps_cost_of_holding.html.

Question 4-6: Discuss whether or not you agree that "holding back the sea" would be the best solution to rising sea level. What other options exist?

Which Areas Will Be Affected?

Sea-level rise will increase flood risks in areas already at or under sea level, like New Orleans, Louisiana, in the United States (already the site of disastrous 2005 and 2012 floods) and coastal Holland in Europe. In Bangladesh, about 17 million people live less than 1 meter above sea level. In Southeast Asia, a number of large cities, including Bangkok, Mumbai, Calcutta, Dhaka, and Manila (each with populations greater than 5 million) are located on coastal lowlands or on river deltas. Particularly sensitive areas in the United States include the states of Florida and Louisiana, coastal cities, and inland cities bordering estuaries.

Moreover, low-lying islands in the Pacific (Marshall, Kiribati, Tuvalu, Tonga, Line, Micronesia, Cook), Atlantic (Antigua, Nevis), and Indian Oceans (Maldives) will be greatly impacted. For example, in the Maldives most of the land is less than 1 meter (3.05 ft) above sea level. A seawall recently built to surround the 450-acre capital atoll of Malè cost the equivalent of twenty years of the Maldive's gross domestic product, according to UN reports.

Question 4-7: In 2009, the U.S. Climate Change Research Program published "Global Climate Change Impacts in the U S."[10] Read and summarize their "Key Findings" section. Which cities are at greatest risk from climate change?

Question 4-8: Summarize the major points of this chapter.

Question 4-9: Discuss the issue of global climate change from the standpoint of sustainability.

[10] http://www.globalchange.gov/publications/reports/scientific-assessments/us-impacts/key-findings.

FOR FURTHER THOUGHT

Question 4-10: Most proposals to reduce the impact of climate change focus on (1) developing more nonpolluting renewable energy sources, (2) adding more nuclear power plants, (3) reducing energy use by conservation, (4) reforestation to "soak up" more CO_2, or (5) removing carbon in fossil fuels before combustion and "sequestering" that carbon in underground reservoirs like aquifers. Some have suggested that societies simply learn to live with climate change.

For now, let us assume that reducing emissions of greenhouse gases becomes a top global priority. After researching this issue, rank the list of options described above in order of decreasing practicality and effectiveness. If you find other options, feel free to include them as well. Good places to find information are the websites of the United Nations Environment Programme, the U.S. Dept of Energy, the U.S. Environmental Protection Agency (www.epa.gov), the National Oceanic and Atmospheric Administration (www.noaa.gov), and environmental organizations such as the Sierra Club and Environmental Defense. Many "think tanks" such as the Rand Corp. (www.rand.org) offer useful reports and commentary. Technology already exists to address the impacts of climate change. The decision to address climate change will be a political one.

Question 4-11: There are mounting concerns that rising atmospheric CO_2 concentrations will cause changes in the ocean's carbonate chemistry system, and that those changes will affect some of the most fundamental biological and geochemical processes of the sea. Go to http://www.pmel.noaa.gov/co2/story/Ocean+Acidification, study the issue, and discuss the impacts of acidifying the oceans.

CHAPTER 5

MOTOR VEHICLES AND THE ENVIRONMENT I

KEY QUESTIONS

- What are the major motor vehicle trends in the United States?
- What are the environmental impacts of motor vehicles?
- What are CAFE standards? Are they working?
- What are hybrids, and what is their long-term significance?
- Are personal motor vehicles compatible with a sustainable society?

BACKGROUND

The model year 2000[1] saw the introduction of the Honda Insight (since discontinued, then reinstated in 2010) and the Toyota Prius (Figure 5-1), the first "hybrid" motor vehicles. These vehicles use gasoline engines, electric motors, and advanced technology to achieve mileages of close to 50 miles per gallon, are also extremely low-polluting vehicles. The Prius was a sales phenomenon. See Table 5-1.

Question 5-1: In what year did hybrid sales peak?

FIGURE 5-1 A 2012 Toyota Prius. More than one million Priuses have been sold in the U.S. since 2000. (Courtesy of Alan Look/Icon SMI/Corbis Images.)

[1] A model year begins on October 1 and ends on September 30.

TABLE 5-1 ■ Sales of Hybrid Vehicles in the U.S.
(Ref: www.hybridcars.com and Statistical Abstract of the U.S., www.census.gov)

Model year	Sales
2000	9,350
2001	20,287
2002	35,000
2003	47,525
2004	88,000
2005	205,749
2006	252,636
2007	352,000
2008	312,000
2009	290,000
2010	274,000
2011	268,000

Question 5-2: Why do you think hybrid sales declined beginning in 2008? [Hint: How did the overall auto market behave?]

With unprecedented nominal (unadjusted for inflation) gas prices exceeding $4/gallon by 2012, the automobile culture, in the view of some observers, is beginning to lose some of its allure, especially among the young. Are we *still* what we drive?

THE ENVIRONMENTAL (AND SOCIAL) COST AF MOTOR VEHICLES

The environmental and social impact of our reliance on internal-combustion-engine-powered motor vehicles is profound. In the United States, these impacts include the following:[2]

- Health-care costs in the billions associated with crashes[3]
- Annual interest cost near $20 billion to finance the purchase of passenger cars
- Hundreds of millions of vertebrates, including thousands of deer, killed each year. Deer crashes are a major source of auto insurance payouts.
- Air and water pollution (see below), including a major contribution to what the Natural Resources Defense Council calls "poison runoff." Poison runoff includes cadmium, zinc, and other heavy metals from tire wear and hydrocarbons from fuel and grease.[4]
- Accelerated urban sprawl as roads are paved
- Fragmentation of habitat by road building, introduction of alien species, etc.

[2] For details, consult websites of these organizations: American Automobile Association, U.S. EPA, Natural Resources Defense Council, Wildlands CPR, Statistical Abstracts.
[3] Insurance Institute for Highway Safety data for 1999 (www.hwysafety.org/safety_facts/safety.htm).
[4] Poison runoff also originates from nonpoint sources (like golf courses) as opposed to point sources such as sewer pipes or factory smoke stacks.

- Dependence on foreign oil from undesirable sources
- Road rage incidents

Here are some additional auto-facts:[5]

- In 2009, there were 10.8 million motor vehicle accidents, resulting in 35,900 deaths.
- In 2009, the average vehicle was driven 11,600 miles.
- In 2009, the U.S. consumed 168.9 billion gallons of fuel for motor vehicles.
- In 2009, urban road congestion cost the average driver $591, each wasting 20 gallons of fuel.
- In 2009, according to the Tax Foundation, the average household spent $216 a month on gasoline and motor oil.

Question 5-3: Do you own a motor vehicle? How similar is the average monthly expenditure on gas and oil to your experience?

WHAT WE DRIVE

Table 5-2 shows sales of motor vehicle by class from 1975 to 2010.

TABLE 5-2 ■ New Motor Vehicle Sales and Leases, 1975–2010, Millions. (Source: Statistical Abstracts of the US)

Year	Car	Light Truck
1975	8.624	2.468
1980	8.979	2.623
1985	11.043	5.193
1990	9.301	5.256
1995	8.635	7.869
2000	8.847	11.529
2005	8.614	12.976
2006	7.821	8.683
2007	7.618	8.842
2008	6.814	6.680
2009	5.456	5.145
2010	5.729	6.044

[5] Statistical Abstracts of the U.S., www.census.gov.

Question 5-4: On the axes below, separately plot sales of each of the above categories for the years 1975, 1985, 1995, 2005, and 2010.

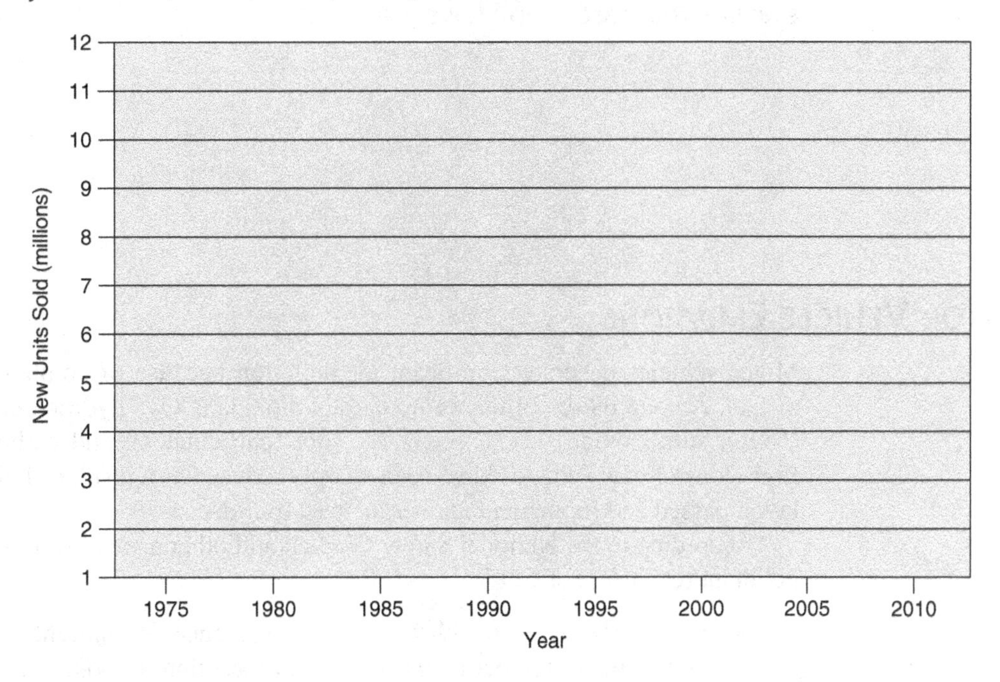

Question 5-5: Calculate the percentage of light trucks (including SUVs) out of total vehicle sales and fill in the table below.

Year	% Light Trucks
1975	
1980	
1985	
1990	
1995	
2000	
2005	
2010	

Question 5-6: Describe trends in sales of light trucks and passenger cars from the data in Table 5-2 and your graph.

One reason that families buy large vehicles like SUVs is laws requiring large car seats for children and infants. Most vehicles simply cannot accommodate more than two of these behemoths.

Question 5-7: List other reasons why people buy SUVs and large trucks. Identify the assumptions you bring to the issue of the "suitability" of these vehicles, and then critically evaluate the reasons you have just listed.

MOTOR VEHICLE EMISSIONS

Motor vehicles generate significant air pollution because of combustion or evaporation of fuel. Among these pollutants are carbon dioxide (CO_2), hydrocarbons, nitrogen oxides (NOx), sulfur oxides (SOx), benzene, soot (particulates), and carbon monoxide (CO). Emissions of new vehicles have been sharply reduced as a result of federal and state legislation passed and implemented over the past four decades.

According to the National Safety Council and other agencies, impacts of each class of pollutant (exclusive of CO_2) are as follows:

- *Hydrocarbons* react with NOx in the presence of sunlight to form ground-level ozone. Higher temperatures enhance the reaction. Exposure can lead to permanent lung damage, among other things.
- NOx contributes to the formation of ozone and acid precipitation.
- SOx contributes to acid precipitation and is toxic at high concentrations. It results from oxidation of sulfur in motor vehicle fuels. In 2006, refiners began to reduce sulfur by as much as 97 percent in diesel fuel. By 2010, diesel for on-road trucks contained only 15 ppm sulfur, according to the U.S. EPA.
- Exposure to benzene, a component of gasoline, can cause cancer and other diseases.
- "Soot" results mainly from diesel vehicles. Particulates are the major source of urban air pollution in many areas.
- CO is a colorless, odorless, lethal gas. CO reduces the flow of oxygen in the bloodstream and can impair mental functions and motor response. In urban areas, motor vehicles generate up to 90 percent of atmospheric CO.

Petroleum combustion, mainly for transportation, accounted for nearly 43% of the United States' annual CO_2 emissions in 2009 according to the U.S. EIA. CO_2 is the major global agent of human-induced climate change. The U.S., in turn, is responsible for 20% of global CO_2 emissions.

Question 5-8: What are environmental and health impacts if emissions are not reduced? Who pays these costs?

Question 5-9: Should people who choose to drive low-mileage vehicles with high emissions pay these costs? How? Be sure to assess your reasoning, using critical thinking principles.

FUEL ECONOMY AND CAFE STANDARDS

The Energy Policy and Conservation Act of 1975 established fuel economy standards for passenger cars. CAFE (Corporate Average Fuel Economy) standards were implemented for 1980 model year cars at 18.0 miles per gallon (mpg). The CAFE standards for cars and light trucks in 2000 were 27.5 and 20.7 mpg, respectively. New standards for the years 2012–2016 will raise average fuel economy for cars to 35.5 mpg and set tailpipe emissions of CO_2 to an average of 250 g/m. For trucks, the standard is 28.8 mpg. In 2011 standards were proposed for 2017–2025, which would gradually raise fuel economy to 54.5 mpg for the 2025 model year. The U.S. EPA estimates that the average cost for compliance will add about $800 to the average cost of a vehicle, but the owner will save about $4000 over the vehicle's life.[6]

Question 5-10: The auto industry often responded to CAFE standards by reducing the weight of some vehicles. How could improving fuel economy this way lead to decreased safety?

Question 5-11: Is the assumption that increasing fuel economy must be achieved only through vehicle "downsizing" valid? Research the issue. What are other ways in which increased fuel economy can be achieved?

Car and Driver magazine conducted an in-depth analysis of the new regulations in 2010 and concluded, "achieving these goals will require various engine and transmission technologies, as well as improved aerodynamics, tires with lower rolling resistance, and materials that reduce weight. After wading through some 1500 pages of documents, we can say that this overhaul of CAFE was carefully considered, involved extensive input from automakers, and—with the new size-based standards—takes into account customer choice in a way that the old system never did. And if gas prices once again head toward $4, customer demand for fuel economy will likely outstrip these regulations."[7]

---

[6] U.S. Environmental Protection Administration, www.ePa.gov.

[7] Csere, Csaba, *Car and Driver*, May 2010.

Question 5-12: Some critics assail CAFE standards for a variety of reasons. Research this issue and evaluate their objections.

For Further Thought

Question 5-13: We discard over 250 million tires a year. Cadmium and other heavy metals partly from tire wear pose a threat to estuaries like San Francisco Bay and Chesapeake Bay. Tire dust also causes human respiratory problems.[8] Evaluate the environmental impact of tire use and discards. How can we reduce tire impact while personal motor vehicle use increases?

Question 5-14: Discuss to what degree low fuel prices are in the best interests of the American consumer.

Question 5-15: How do low fuel prices affect the nation's security?

Question 5-16: How do low fuel prices affect the economics of substitute fuels like biodiesel or ethanol?

Question 5-17: How do low fuel prices affect decisions by energy companies to seek new supplies?

Question 5-18: In 2004 Ford Motor Co. introduced the Escape hybrid SUV, with mileage ratings of 36 mpg "city," 31 mpg "highway." Research whether sales of the hybrid version have been successful in the United States. Interpret your findings.

Question 5-19: By 2012, several manufacturers were again selling all-electric models, including Nissan and General Motors, and more were planning electrics. Research the advantages and disadvantages of all-electric vehicles.

Question 5-20: Research the impact that battery technology has on the cost and range of electric vehicles.

[8] *Rachel's Environment and Health Weekly.* 1995. Tire Dust. #439.

CHAPTER 6

AQUATIC ENVIRONMENTS

FIGURE 6-1 Threatened red-legged frogs (*Rana aurora draytonii*) depend on shrinking aquatic habitats, which have been reduced in quality and area in California.

INTRODUCTION

California red-legged frogs (*Rana aurora draytonii*) were once familiar aquatic animals in the Western United States (Figure 6.1). Mark Twain wrote about these frogs, and about the California gold miners who gambled on their leaping ability in "The Celebrated Jumping Frog of Calaveras County." Larger than any other native frog in the region, these amphibians were intensively harvested for food in the 1800s. Overharvesting, along with widespread diversion of surface water for irrigation, pollution of streams, and introduction of aggressive competitors such as the bullfrog, caused a steady decline in red-legged frog populations over the past century, according to the U.S. Fish and Wildlife Service (2001).

Now listed as a threatened species, red-legged frogs are emblematic of our general concerns about aquatic ecosystem health.

Amphibians all over the world are declining, not for a single reason, but because their fragile wetland habitats, and the pure water they require, are impacted by so many different human activities. Synthetic hormones released as pollutants in wastewater, nutrients and sediments from agricultural runoff, draining of marshlands, exotic diseases, increased ultraviolet penetration of atmospheric ozone, and gradual changes in climate have all been suggested as possible contributing factors (Davidson et al., 2001). Although rare species like the red-legged frog do eventually merit protection under the Endangered Species Act, recovery efforts are daunting after the species has been reduced to a few fragmented populations. The Endangered Species Act has been a valuable conservation tool, but its application usually involves damage control rather than prevention. Basing conservation policy solely on a species' endangered status is like waiting to see the dentist until you have only one tooth left in your mouth. Ecologists recognize that effective biodiversity protection for wetland and aquatic species depends on ecosystem-level understanding of our environment, and thoughtful management of entire watersheds. Mark Twain's miners may have bet on one frog at a time, but our task is to improve the odds for the aquatic community as a whole.

What are the most important environmental factors affecting life in aquatic systems? Dissolved oxygen is certainly critical, and this variable is always linked with temperature. Cold water is able to carry more oxygen than warm water. At one atmosphere of pressure, a liter of water near the freezing point can dissolve about 14 milligrams of oxygen. The same water heated to 30° C can dissolve only about 7 milligrams of oxygen. In other words, water in a temperate zone creek loses nearly half its ability to carry dissolved oxygen between January and July. Metabolic activity by bacteria or by aquatic organisms reduces oxygen levels below this maximum limit, so accelerated decomposition of organic matter in warm weather can make this difference even more extreme. Add to this the effects of water movement, with stream flow slower and more sluggish in summer, and faster with more aeration in winter, and the difference is greater still.

FIGURE 6-2 Trees reduce water temperature in shaded upper reaches of a stream.

Within a watershed, a stream tends to be more oxygenated in its upper reaches, where flow rates are greater and overarching trees shade the water (Figure 6.2). As a stream slows down and gets wider in its lower reaches, temperature increases and oxygen levels decline. A slower flow rate downstream provides less mixing, and this reduces oxygen levels as well. These changes in the physical environment affect life forms in the stream. Active fish with high oxygen demand are limited to cool streams with rapidly flowing water (Figure 6.3). Fish inhabiting warm stagnant waters tend to lower their metabolic demand for oxygen by moving slowly and spending much of their lives sitting still and waiting for prey to come to them (Figure 6.4). Human activities that increase water temperature, such as cutting shade trees away from stream banks or discharging heated water from power plants into rivers, significantly affect resident aquatic life.

FIGURE 6-3 Salmon migrate upstream to spawn in cool, oxygen-rich waters.

CHECK YOUR PROGRESS:

Aside from the obvious advantage of leaving more time for tadpole development, why might so many species of frogs in temperate zones shed their eggs very early in springtime?

Hint: Frog eggs need high levels of dissolved oxygen for development.

FIGURE 6-4 Brown bullhead catfish tolerates warm waters with low oxygen.

In a lake, warm water floats on top of cold water, so **stratification** of the water column occurs as the sun warms the surface (Figure 6.5). The warm upper layer, called the **epilimnion**, is the most biologically active. Since light penetrates only a few meters in most freshwater lakes and ponds, photosynthesis is limited to the epilimnion. The cooler, darker zone at the bottom is called the **hypolimnion**. Algae and other plankton "rain" down from the epilimnion above, to be decomposed by bacteria living in the mud at the bottom. There is little mixing of water between layers, so oxygen used up by these decomposers is not immediately replaced. Over the summer, the hypolimnion becomes richer in dissolved nutrients, but poorer in dissolved oxygen. Growth of algae near the surface slows down in midsummer, because their recycled nutrients are trapped in the hypolimnion.

On the interface between the warm epilimnion and the cold hypolimnion is a dividing line called the **thermocline**. This boundary layer can be found near the surface in early summer. If you have ever been swimming in a northern lake in June, you know the "toe test" may indicate warmth near the surface, but the near-freezing hypolimnion becomes all too apparent when you jump in. Through the summer, the thermocline drops lower and lower as solar heating expands the epilimnion. When fall comes, stratification is reversed.

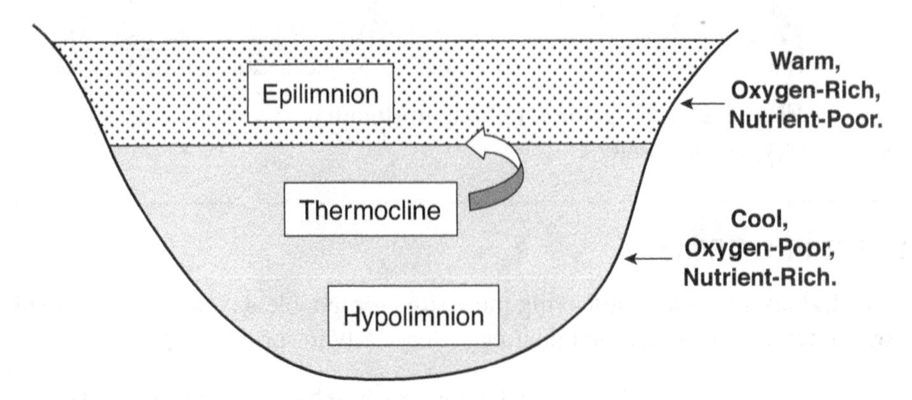

FIGURE 6-5 Temperature stratification in a freshwater lake.

Water exposed to cold air gets colder and heavier, so surface water sinks to the bottom. This mixes the upper and lower layers of the lake, a phenomenon known as "**fall turnover**." The thermocline disappears as epilimnion and hypolimnion are blended together. Since water resists temperature changes, it can take a long time to cool the entire lake to 4° C, which is the temperature at which water is heaviest. Below that temperature, water molecules begin organizing themselves into the lattice of crystalline ice, becoming less dense than liquid water. By the time ice forms on the top of a lake, the entire water column has been cooled to 4° C, all the while mixing nutrient-rich waters from below with oxygen-rich waters from above. By the time of the spring thaw, surface waters can again support algae growth as the epilimnion begins to re-form in a new season.

CHECK YOUR PROGRESS:

If a shallow lake and a deep lake both have the same surface area and the same average water temperature of 20° C, which will freeze over first when winter comes? Why?

Answer: Deep lakes take longer to freeze over because all the water must be cooled to 4° C before any of it turns to ice

In addition to temperature and dissolved oxygen, light is a critical variable in aquatic systems. Aquatic food chains are built on productivity of microscopic algae, called **phytoplankton**, floating in the water. These algae tend to float near the surface because light intensity declines exponentially with depth. For example, if we measure red light of wavelength 620 nm at 100% intensity at the surface of a perfectly clear lake, its intensity declines to 10% at a depth of 9 m, 1% at 18 m, and 0.1% at 27 m. Not all wavelengths of light penetrate water equally well. Blue light, with wavelengths in the 400–500 nm range, is transmitted much more effectively than red light (600–700 nm). When sunlight composed of all colors enters water, the red wavelengths are preferentially absorbed, so light that has passed through water looks blue. Chlorophyll, the photosynthetic pigment for land plants and many algae, absorbs blue and red portions of the spectrum, so the rapid attenuation of red light near the surface significantly reduces energy available to phytoplankton. In deepwater marine habitats, some algae actually use a different photosynthetic pigment, which absorbs only the blue light available at that depth. Since their cells reflect red light rather than absorb it, they have a red appearance when brought up to the surface. They are appropriately called red algae.

Suspended particles and pigments interfere with light transmission in direct proportion to their concentration. This is the principle behind the spectrophotometer, which you may have used in chemistry laboratory to measure the concentration of a colored solution. Silt and clay soil washing into streams reduces photosynthesis by interfering with light transmission to algae and submerged plants. This reduction in clarity is called **turbidity**. It can be measured in an instrument similar to a spectrophotometer, called a turbidimeter. Standard units of turbidity, called nephalometric turbidity units, or N.T.U., are often included in water quality assessments of streams. A more traditional way to measure water transparency is with a Secchi disc (Figure 6.6). The Secchi disc is a round flat piece of metal or plastic, 20 cm in diameter, painted white and black in alternating quarters and weighted so that it will sink. A ring in the center is tied to a chain or rope which is marked off in meters, and the Secchi disc is lowered over the side of a boat. The observer lets out the line until the Secchi disc disappears from view. Then the observer pulls up the line until the Secchi disc is just visible, and a depth measurement is recorded. Ideally, Secchi disc readings are taken between 10:00 am and 2:00 pm, and observation is off the shady side of the boat for better visibility. The greater the depth a Secchi disc can be seen, the clearer the water.

CHECK YOUR PROGRESS:

What factors determine the amount of light energy available to an algal cell suspended in a lake?

Answer: Colors of light needed for photosynthesis, distance from the surface, and turbidity of the water

Finally, dissolved chemicals determine viability of aquatic life. Nutrients such as phosphates, nitrates, and potassium are needed in small amounts for biosynthesis, but can wreak havoc with aquatic communities if added in large amounts. A body of water receiving too much fertilizer is said to be **eutrophic**, which literally means "overfed." Algae grow exponentially in "blooms" or floating mats that block out all the light below the surface. Shaded algae die and decompose. Bacterial decomposition uses up dissolved oxygen, with predictable results for other aquatic organisms.

Some sources of pollution are easier to find than others. **Point sources**, such as industrial spills, pulp mill effluent, and insufficiently treated municipal wastewater, enter a body of water at an identifiable location. **Non-point sources**, which include runoff from lawns or farm fields, acid rain, manure from feed lots, silt from road construction, or leachates from mining operations, are more diffuse in their origin and thus harder to identify and control. Toxicity may be acute, resulting in dramatic fish kills, or chronic, causing gradually declining biodiversity in an affected waterway. The growing number of biologically active compounds entering streams from antibiotics in animal feed and from medicines incompletely metabolized by humans has more recently raised concerns about effects on animal fertility, reproduction, and development.

Chemical tests have been designed for many kinds of pollutants in water, but it is difficult to assess how much damage is being inflicted on stream biota in this way. A more direct approach is to test the water directly on organisms, monitoring their viability over time. So called **bioindicator** organisms are chosen for their short life spans and high sensitivity to pollutants. Two organisms used routinely for biomonitoring work are *Daphnia* (also called water fleas—Figure 6.7) and the small fish *Pimephales promelas* (also called fathead minnows—Figure 6.8). The premise of this biological approach is that water clean

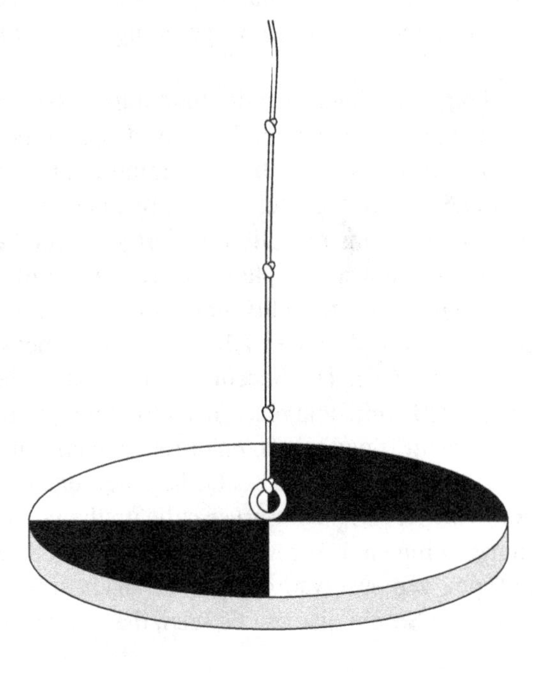

FIGURE 6-6 A Secchi disc is lowered into a body of water from a boat to measure turbidity of the water.

enough to support bioindicator organisms is probably safe for the aquatic ecosystem as a whole. Bioindicators are studied in two settings. Laboratory experiments expose a subject population to suspected toxins or suspect water sources in carefully controlled experiments. For example, a pesticide might be screened by making up a series of dilutions and exposing fathead minnows in aquaria to these dilutions over a period of time. A concentration of the pesticide just strong enough to kill half of the fish is called the lethal dose for 50%, or **LD_{50}** value. This critical concentration can then be compared with LD_{50} values for other pesticides to determine which has the lowest effect on fish if it runs off of fields into streams.

A second biomonitoring approach looks at organisms already living in a stream. A convenient group of bioindicators for on-site assessment are **benthic macroinvertebrates**. Benthic means living on the bottom, and macroinvertebrates are non-vertebrate animals large enough to be seen with the naked eye. This group includes strictly aquatic species, like amphipods and snails, that live all their lives in the water. It also includes the aquatic larvae of many insects, such as dragonflies, that live near aquatic habitats as adults and lay eggs in the water (Figure 6.9). Some of these macroinvertebrates, including the juvenile forms of mayflies, caddis flies, and stoneflies, are quite sensitive to pollution. If any toxins have entered the stream in the past few months, these organisms will be absent from the benthic community. Other macroinver-tebrates, including aquatic annelids and the larvae of midges, can tolerate a high pollution load. By comparing numbers of pollution-sensitive vs. pollution-tolerant species, ecologists can develop a water quality index that is more inclusive than a battery of chemical tests. In a sense, the bioindicator species have been monitoring pollution in the stream 24 hours a day, 365 days a year, for every pollutant that can harm organisms, right up to the time of your arrival.

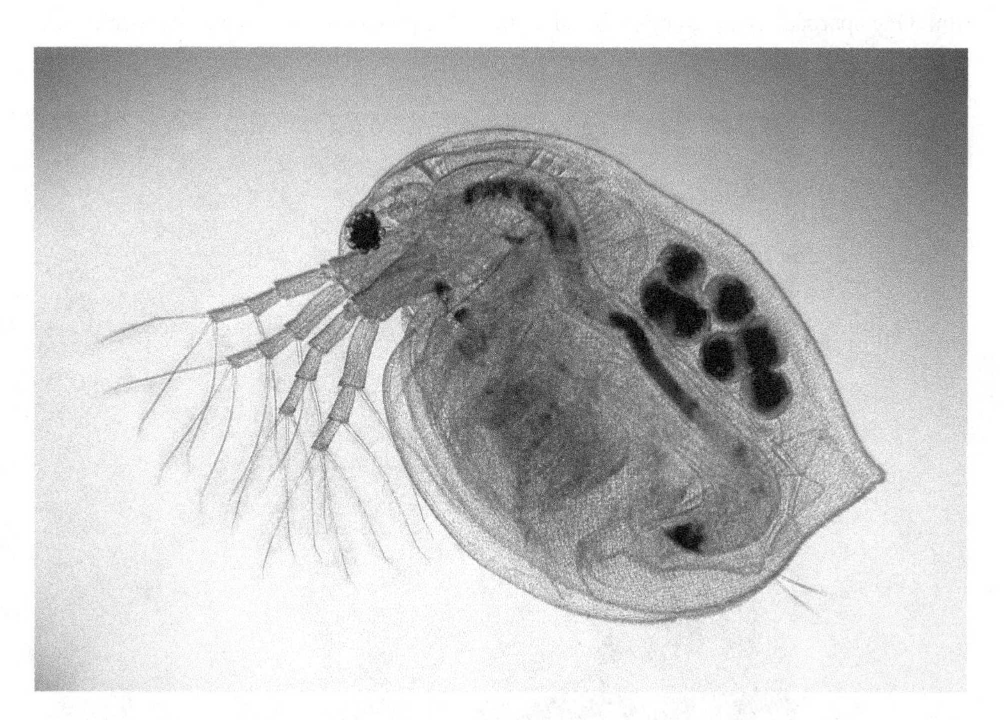

FIGURE 6-7 Daphnia are commonly used bioindicators for pollution in aquatic environments.

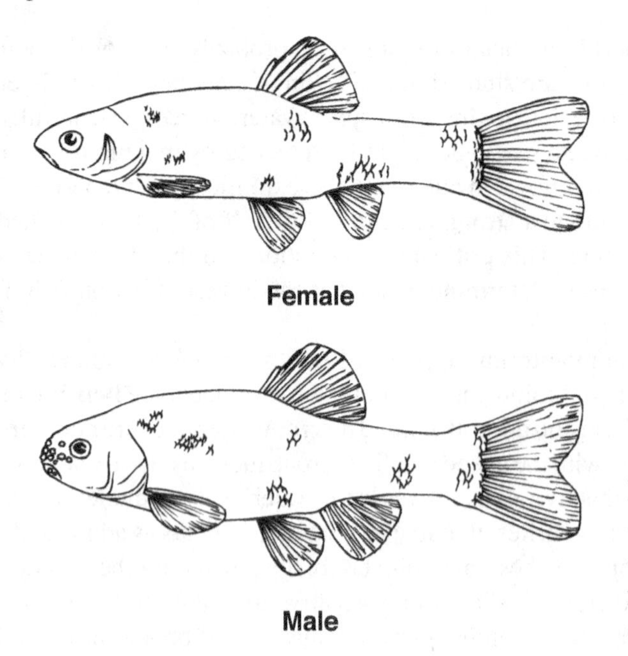

FIGURE 6-8 Fathead minnows (*Pimephales, promelas*) are commonly used as bioindicators for toxins affecting survival and development of vertebrate animals.

CHECK YOUR PROGRESS:

To monitor non-point sources of pollution, what advantages would you see in a series of chemical tests, such as pH, ammonia nitrogen, and dissolved oxygen assays? What advantages would you see in using bioindicators instead?

Hint: One approach is more comprehensive, the other potentially yields a more specific diagnosis of the problem.

FIGURE 6-9 A dragonfly deposits eggs on aquatic vegetation.

METHOD A: DISSOLVED OXYGEN AND TEMPERATURE

[Laboratory/outdoor activity]

RESEARCH QUESTION

How does temperature affect dissolved oxygen concentration?

PREPARATION

A series of five water baths at different temperatures between 0° C and 40° C will be needed for the laboratory portion of this experiment. If you do not have water baths that can be set at a range of temperatures, a fair substitution can be made with inexpensive fish tank heaters placed in 1000-ml beakers. Beakers can also be left overnight in an incubator or refrigerator. Make sure the water stays uncovered at temperature for several hours so that oxygen has a chance to equilibrate before testing.

Kits that include a chemical test for dissolved oxygen are inexpensive and widely available. These are adequate for the experiment if groups cooperate on testing and share data to save time. A portable meter with a dissolved oxygen probe is a better way for an individual to collect all needed data in a short time. For greater accuracy, groups can be instructed to pool their data, and to plot a standard curve through mean values for the class.

Outside water sources depend on your campus environment. Fish ponds, fountains, puddles, ditches, and adjacent streams are all possible sample sites. If outdoor sources do not exist, try measuring dissolved oxygen in aquaria as a model system.

MATERIALS (PER LABORATORY TEAM)

Access to five water baths, of varying temperature

Access to at least one outdoor water source

Dissolved oxygen test kit or portable dissolved oxygen meter

Thermometer for assessing water temperature

PROCEDURE

1. In each of the water baths, measure temperature and dissolved oxygen. Record your results in the *Data Table for Method A*.
2. Generate a graph on the *Results for Method A* page, showing the relationship between temperature (x-axis) and dissolved oxygen (y-axis). Draw a smooth curve through your points. This will be your standard curve.
3. Find a water source outdoors (or use an aquarium if necessary). Measure temperature and dissolved oxygen. Note water clarity, organic debris, flow rate, any organisms present, and other factors that may affect dissolved oxygen concentration. Repeat for more than one site if you can.
4. Draw circled points representing your field or aquarium data on the graph with your standard curve. Use your observations and measurements to answer *Questions for Method A*.

DATA TABLE FOR METHOD A

PURE WATER SAMPLE NUMBER	TEMPERATURE °C	DISSOLVED OXYGEN (mg/l)
1		
2		
3		
4		
5		
FIELD SAMPLE(S) (Describe sites below.)	**TEMPERATURE °C**	**DISSOLVED OXYGEN (mg/l)**

RESULTS FOR METHOD A:

Dissolved Oxygen as a Function of Water Temperature

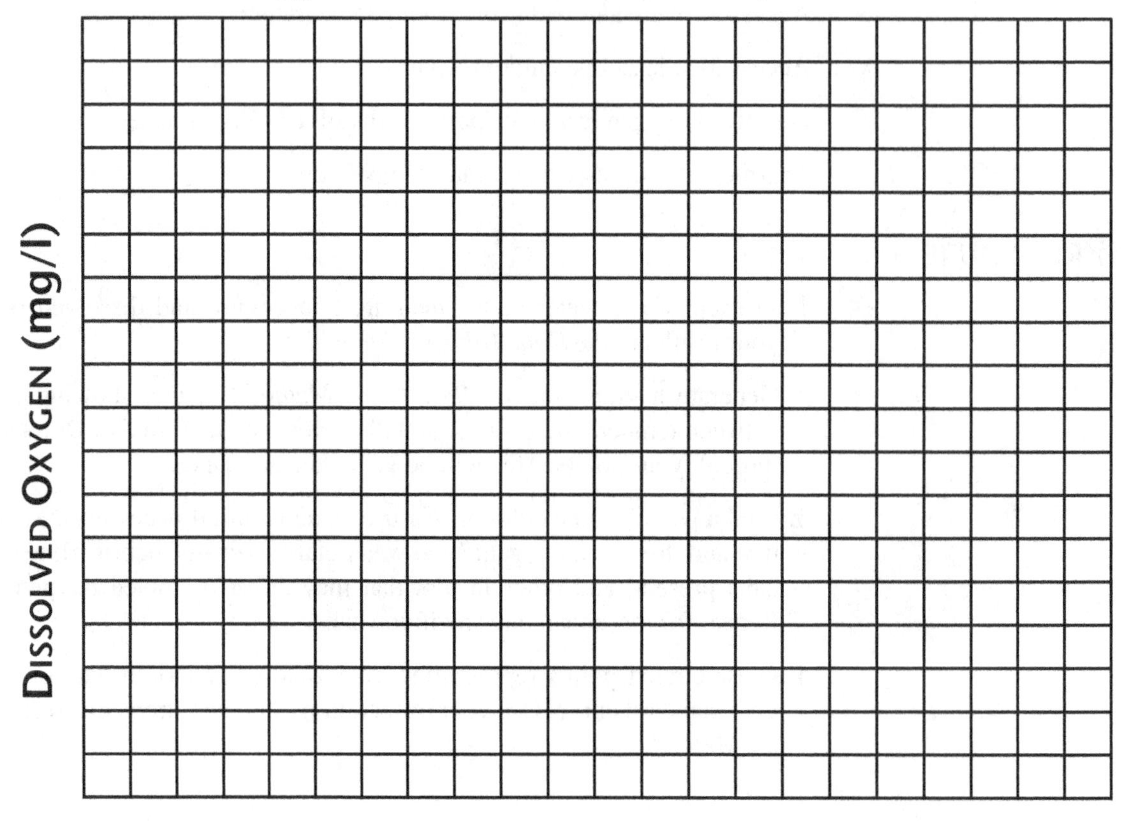

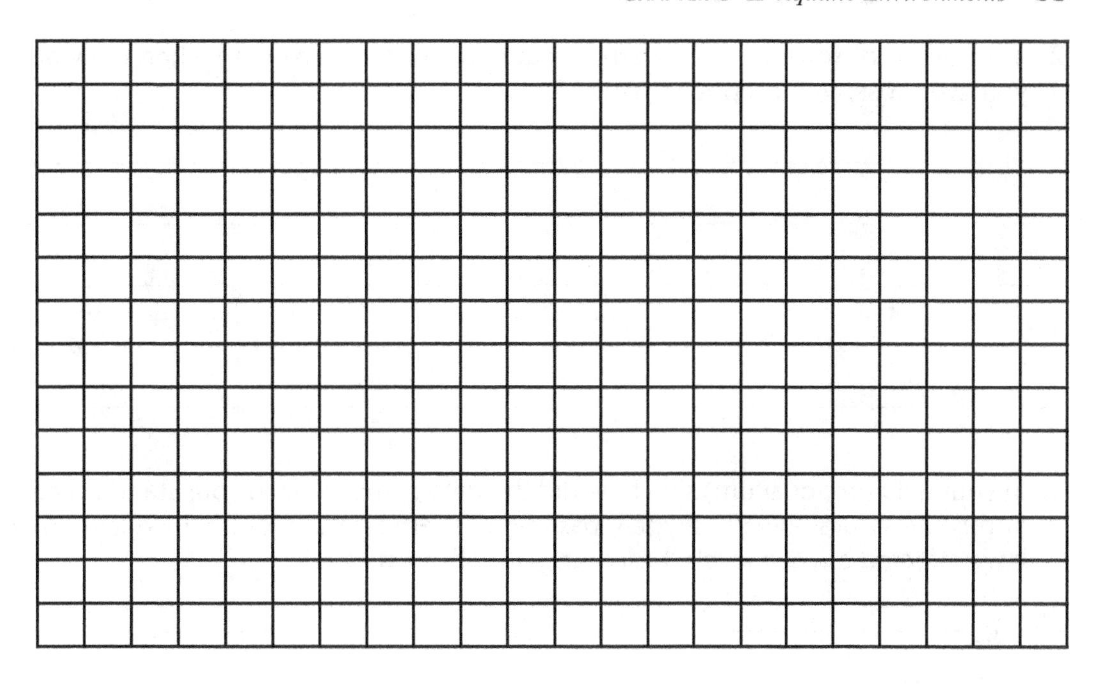

WATER TEMPERATURE (°C)

FIELD OBSERVATIONS:

QUESTIONS FOR METHOD A

1. Describe the shape of your standard curve. Is this relationship linear or curvilinear? Explain.

2. If oxygen is dissolved in cold water, what happens to the oxygen when the water warms up? Have you ever observed this phenomenon?

3. In your field (or aquarium) site, how did the data compare with your standard curve? Was the water at this site saturated with oxygen, based on its temperature, or not? Explain other factors that may have affected oxygen levels, based on your observations.

4. How might the relationship between temperature and dissolved oxygen explain the adaptive significance of symbiotic algae within the bodies of coral polyps on tropical reefs?

5. Trapping of infrared radiation by carbon dioxide in the atmosphere has the potential to warm the atmosphere a few degrees over the next century. This does not seem very significant to a terrestrial mammal like yourself, but how might this change affect tropical aquatic habitats?

CHAPTER 7

Global Water Supplies: Are They Sustainable?

KEY QUESTIONS

- How much water do people need?
- How abundant are world supplies of clean water?
- How much water is used in agriculture?
- Who provides the world's water?
- What effects will population growth and development have on water supplies?
- How could disagreements over water lead to conflicts among nations?
- What effects will climate change have on water supplies?

INTRODUCTION

In 1977, California was in the midst of a powerful drought. Residents were forbidden to water their lawns or wash their cars. Children learned a rule governing toilet flushing, "If it's yellow let it mellow." One of the greatest hardships the state's middle-class residents faced was a prohibition against washing driveways—a popular weekend activity. Springing to the rescue, entrepreneurs marketed a heretofore rarely used Japanese invention—the leaf blower. By 1980, the leaf blower had found its way into California's (and America's) heart and all because of a water shortage. It is now one of the top sources of air pollution in the state.

Population growth and looming climate change threaten to alter weather patterns and could materially reduce California's access to fresh water. Is this a water crisis?

Half a globe away, Bangladesh faces another kind of crisis: arsenic contamination of the shallow aquifers on which many Bangladeshis rely. Health officials have shown that cholera, an often deadly bacterial infection, can be controlled by a simple and cheap cloth-filtration system for drinking water, but arsenic cannot be so easily removed. Is this a water crisis?

GLOBAL WATER USE

Throughout the 20th century, while human population tripled, water use increased by a factor of six, according to the World Water Council.[1] Population is projected to increase by 30–40% by 2050. How will this affect global supplies, and global water demand?

Global water use is shown in Figure 7-1.

Question 7-1: Which economic sector—agriculture, industry, or municipalities—uses the most water?

[1] World Water Council, www.worldwatercouncil.org/index.php?id=25.

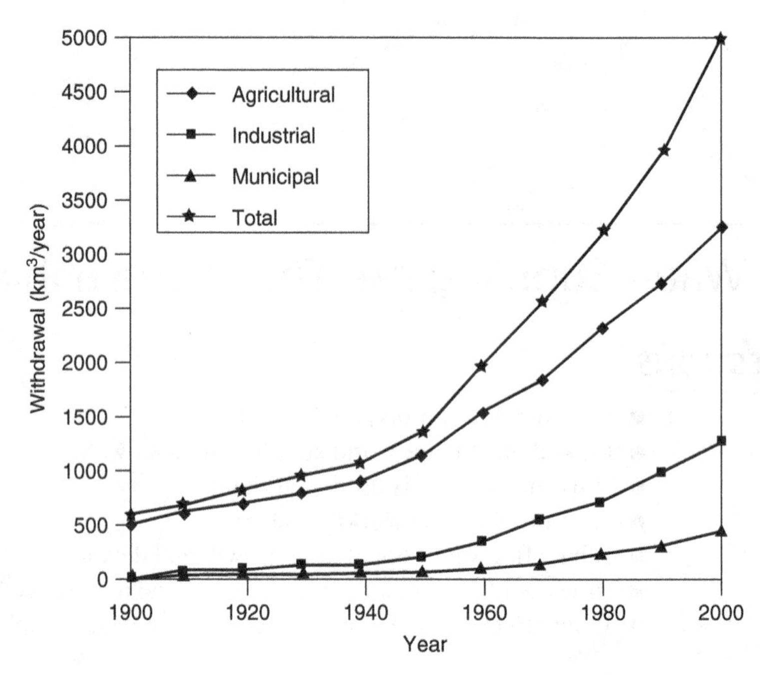

FIGURE 7-1 Global water use by sector. (Worldwatch Institute, *Water in Crisis,* www.worldwatch.org. Courtesy of World Watch Magazine.)

Nearly one in six of the planet's 7.0 billion people presently lack access to safe water, and at least 2.4 billion do not have access to adequate sanitation. Nearly 1.5 million children die needlessly each year because of preventable water-borne diseases. Water demand is projected to grow by as much as 40 percent over the next twenty years. If this issue is to be addressed, hundreds of billions of dollars must be invested globally in water infrastructure over the next five decades.

We begin investigating global water issues with a review of the hydrologic cycle.

The Hydrologic Cycle

Water at the Earth's surface moves through three *states*—liquid, solid, and vapor—and is carried from place to place at the surface of the Earth. The movement of water at the Earth's surface is called the *water,* or *hydrologic, cycle.* Figure 7-2 shows the water cycle. The labeled components of the cycle are *evaporation, precipitation, runoff,* and *infiltration.* Water that infiltrates into pores in soil, sediment, or rock is called *groundwater,* where it may be stored for millennia. Water may similarly be stored for long periods as ice. Table 7-1 shows where the water is found.

Question 7-2: Where is most fresh water at the Earth's surface?

Question 7-3: How could climate change affect the distribution of water in Table 7-1?

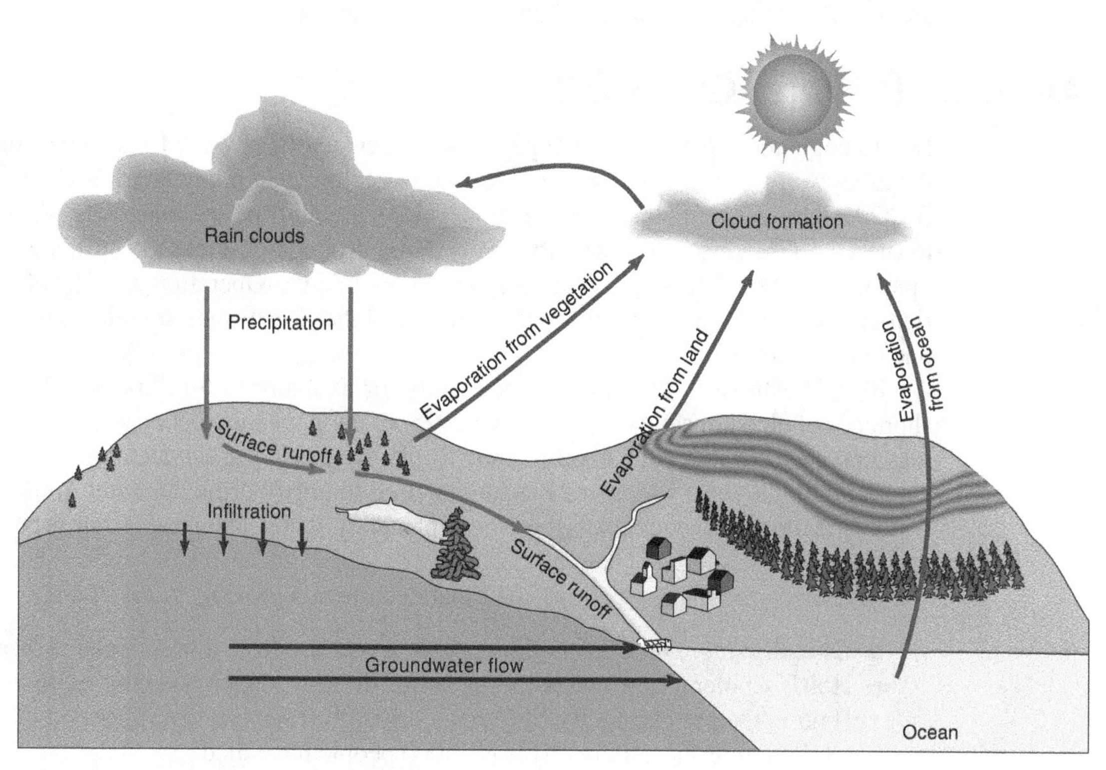

FIGURE 7-2 The hydrologic cycle. (Keller, E. A. 2006. Natural Hazards. Pearson Prentice Hall, Upper Saddle River, NJ. Courtesy of E. A. Keller/Pearson Education.)

TABLE 7-1 ■ Where Is the Water?

	As a Percent of All Water	As a Percent of Global Fresh Water
Earth's Oceans	97.50%	—
Glaciers & Permanent Snow	1.74%	68.70%
Fresh Groundwater	0.76%	30.10%
Salt Groundwater	0.99%	
Ground Ice & Permafrost	0.02%	0.86%
Fresh Water Lakes	0.007%	0.25%
Salt Water Lakes	0.006%	
Soil Moisture	0.001%	0.05%
Atmosphere	0.001%	0.04%
Swamps	0.0008%	0.03%
Rivers	0.0002%	0.006%
Plants	0.0001%	0.003%

HOW MUCH IS WATER WORTH?

One observer said that while we might think that diamonds are more valuable than water, it really depends on how thirsty we are. Most things that have economic value have substitutes to which users can turn if the price becomes too high: natural gas for oil, mass transit for autos, spinach for broccoli, etc. But there is no substitute for water. Moreover, the demand for water is growing with human population and, locally, with affluence. Until recently, most people in wealthy countries thought little about the ecological, economic, and social consequences arising from a lack of clean water.

HOW AVAILABLE IS WATER GLOBALLY?

Due to population increase, global per capita water supplies by 2011 were one-third lower than they were in 1970, and water *quality* was declining in many areas, according to the IPCC.[2] Eighty percent of all illnesses in developing countries are water-related, according to the UN. Growing water scarcity has become a major obstacle to "sustainable development." A 2002 OECD (Organisation for Economic Co-Operation and Development: a rich-nations think-tank) report[3] concluded that global freshwater supplies must grow by one third by 2020.

By UN estimates, two-thirds of humanity, or as many as 5 billion people, will face shortages of clean freshwater by the year 2025. Even in wealthy countries, water problems may become serious. Some toxic organisms like *Cryptosporidium* are already resistant to chlorination, the most widespread technique used to purify drinking water. In the view of many hydrologists, better management of planetary water resources is an urgent global need.

Here are some snapshots from 2012 that document a growing water crisis:[4]

- **Globally**: more than 3.5 million people a year die from a water-related disease.
- **Haiti:** Cholera, a water-borne bacterial disease began ravaging earthquake-torn Haiti in October 2010. By May 2012, 140,000 cases had been reported in the capital, Port au Prince, alone. At least 7,000 people have died.
- **China:** Beijing's 20 million residents have faced increasing water shortages since 1999. By 2011, per capita water supplies had fallen to 100 cubic meters a year, one-twentieth of the national average. Groundwater was down by 37% and surface supplies by 59%.[5]

Are Water Conflicts Ahead?

On the global stage, at least 260 rivers—the Danube, the Volga, the Ganges, the Brahmaputra, and the Mekong, to name a few—flow through two or more countries. Likewise, neither lake basins nor groundwater aquifers recognize national boundaries. Three-fifths of the world's population lives in the watersheds of international freshwater systems. No global doctrine governs the allocation and use of these international water bodies. Even in the United States, water doctrines and laws vary from state to state. We revisit this subject in "For Further Thought" at the end of the chapter.

WHO COLLECTS AND DISTRIBUTES THE PLANET'S WATER?

Sizeable investments are required to extract, purify, and distribute water. Should economic and population forecasts materialize, the UN estimates that at least US $180 billion will be required globally to expand supplies over the next two decades. This estimate does *not* include what will be required to rehabilitate or modernize existing systems, nor does it include operation and maintenance costs.

---

[2] Intergovernmental Panel on Climate Change. See for example www.ipcc.ch/pdf/assessment-report/ar4/wg1/ar4-wg1-spm.pdf.

[3] OECD, http://www.oecd-ilibrary.org/economics/oecd-annual-report-2002_annrep-2002-en.

[4] www.water.org/resources/headlines.htm.

[5] *China Times*, 3/15/2012.

There are a number of points you need to know to better understand water issues.[6]

1. First and foremost, *waste* is endemic in the system. Leaking pipes, evaporation from reservoirs, inefficient irrigation and household practices, as well as pollution, all waste water on a vast scale. For example, a substantial part of New York City's water supply disappears unmetered, presumably through leaking underground pipes. And unlined, uncovered irrigation canals can lose more than a quarter of their water through evaporation and seepage.
2. The benefits of enough fresh water extend far beyond anything measurable by economics. For example, were lack of water or polluted supplies to generate large numbers of refugees, the costs would be borne by other segments of society or other nations.
3. Water distribution systems are natural *monopolies,* since it is impractical to have numerous, competing irrigation systems or public water supply systems.

Question 7-4: Does our statement in #3 above violate American "free market" principles? Is it important to force competition on the global water system? Discuss and cite evidence for your position.

4. The pricing of water is usually inefficient. Governments typically build and operate water distribution systems, and the prices charged are often not market-based. Charging a price for water that includes all costs is politically difficult. Politicians in wealthy democracies are often fearful of offending large-scale water users, like irrigation districts in California. And riots may occur in poor countries if water prices are raised.
5. Finally, although water circulates globally, its use by humans is local. Thus, water problems are best dealt with on the regional to local scale. However, national and international agreements are an essential framework for local and regional solutions. "All of us live downstream" means that, since water circulates globally, we are all vulnerable to the effects of water pollution.

IMPACTS OF CONTAMINATED WATER

According to the UN, water scarcity is one of the major factors driving mass migration, creating increasing numbers of human refugees.[7] Water pollution impacts human health in three main ways.

- *First,* humans need access to a minimum amount of clean water—at least 50 cubic meters a year—both to ensure physical survival and to meet minimal hygiene demands.

[6] OECD Factbook, 2010, www.oecd.org/.

[7] United Nations, http://www.un.org/waterforlifedecade/scarcity.shtml.

Question 7-5: How does this number compare to the water available to the average resident in China's capital, cited above?

Question 7-6: How much water is used in the average shower, or to flush a toilet? (For comparison, the average person in the United States uses about 180 gallons per day—about 680 L).

- *Second,* drinking or bathing in water containing animal or human waste facilitates transmission and proliferation of disease-bearing organisms. The most common waterborne infectious and parasitic diseases include hepatitis A, cholera, typhoid, roundworm, guinea worm, leptospirosis, and schistosomiasis. In developing countries, according to the World Bank,[8] diarrheal diseases alone cause an estimated 3 million deaths and 900 million episodes of illness annually, mainly affecting children. We will revisit this topic in "For Further Thought" at the end of this chapter.
- *Third,* surface water and groundwater can dissolve and/or transport inorganic and organic chemicals, heavy metals, and other toxins. These can cause illness, cancer, birth defects and other mutations and can impair immune system function as a result of direct (drinking contaminated water) or indirect (eating plants or animals harvested from contaminated water) exposure.

WATER USE IN THE UNITED STATES

Figure 7-3 shows water use in the United States through 2005, the last year for which data were available when we went to press. The United States uses about 410 billion gallons each day. This figure includes both *consumptive* use (the water is not put back where it came from) and nonconsumptive use (the water is put back after use).

Question 7-7: Describe changes in U.S. per capita water use over the period 1975–2005.

[8] World Bank, World Development Report, www.worldbank.org.

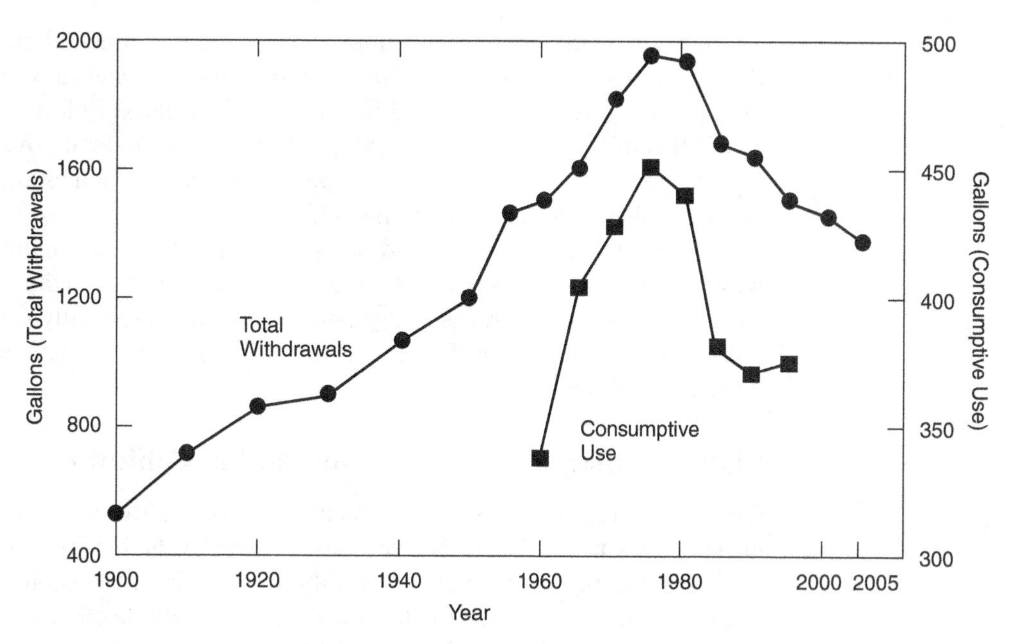

FIGURE 7-3 U.S. water use to 2005, per capita, per day. *Consumptive use* refers to water that is removed and not returned to a system, e.g. agricultural and manufacturing uses.

Focus: Water Use in California

In the United States, water plays a critical role in nearly every state west of the Mississippi River. California is a great example. Most of the state has a Mediterranean or desert climate, and rain falls mainly between October and April.

California is home to 38 million people and adding upwards of 300,000 new residents a year. Up to three-fourths of the water used in California, or about 34 million acre-feet, is by agriculture, even though agriculture provides only about 4 percent of the state's GDP and 1 percent of its jobs. Water has been heavily subsidized to agribusinesses for decades. In parts of the Central Valley, 1 acre-foot (325,000 gal. or about 1228 m^3) could be purchased in 2005 for about $7.50 (a rate of about 770 L for a penny). At the same time, farmers in San Diego County were paying $400 per acre-foot, and residents of Santa Barbara faced a charge of $1,900 per acre-foot.

Question 7-8: At $1900 per acre-foot (325,000 gallons) how many gallons could be bought for a dollar?

Until recently, in many California cities, water was viewed as "too cheap to meter." Even during droughts, users paid a flat fee meaning conservation couldn't be legislated.

Subsidized prices to agribusiness meant massive waste. Most irrigation ditches were made of unlined dirt, which meant that up to a quarter of the water shipped from California's rivers to farm fields trickled into the soil before reaching its destination. Up to 25 percent more evaporated from ditches in 100-degree heat or from reservoirs behind vast dams.

Subsidized prices for water imposed heavy environmental costs on wildlife as well. Rivers like the San Joaquin routinely ran dry, and seasonal runs of salmon and steelhead disappeared, forcing commercial fishers out of business. Before passage of the Omnibus Water Bill in 1992, much of the water supplied by the federal government, at subsidies of up to 90 percent, was used to grow surplus crops like hay and alfalfa; in other words, crops for which there was insufficient market.

The next twenty years could see precedence-shattering innovation to reduce water waste and to restore ecosystems throughout the state. But even those measures might not be sufficient. Computer forecasts suggest the West, and especially California, could become more drought-ridden over the next century, similar to the climate of the prehistoric past as deduced from tree rings.

Climate Change and Water Demand in California

Climate change could have a profound effect on California's water supplies. Studies of ancient trees in the Sierra Nevada have yielded valuable insight into climatic variability in California over the past several thousand years. Bristlecone pines, the oldest living trees, have rings by which the trees can be accurately dated, and some of them are 4,000 years old. Such studies confirm that California's climate over the past millennium has been extremely variable, with severe drought conditions persisting for decades to centuries.

We can now incorporate this historical record into computer forecasts of future climates based on changing levels of greenhouse gases. California's climate is "forecast" (see Chapters 3 and 4) to become more extreme than in the recent past. Should such forecasts, or something even more severe, become the norm, these climatic extremes will critically stress the state's water storage and delivery system and may force agonizing choices between protecting aquatic ecosystems and providing the massive amounts of water consumed by industrial agriculture, commerce, and municipalities.

Question 7-9: Study Table 7-2. How does California compare to the other cities in the U.S.? To Australia? How does urban use compare to the minimal human water requirement of 50,000 liters per year (1 gal. = 3.8l)?

Question 7-10: California's population could top 50 million by mid-century. Discuss the implications for sustainable water use under this scenario.

TABLE 7-2 ■ Water Use in the Western US and Australian Cities (Units are gallons per capita per day [GCPD])

Location	Residential Use** (gpcd)	Urban Use** (gpcd)
Portland, OR	60	116
Albuquerque, NM	74	154
Tucson, AZ	97	144
Denver, CO	104	160
California	**111**	**162**
San Francisco	54–56*	95–102*
Los Angeles	77*–107	139*–154
San Diego	79*–113	136*–157
Oakland/East Bay	87*–100	138*–146
San Jose	91*–97	156–160*
Sacramento	93–128*	142–247*
Australia	**63**	**100**
Melbourne	53	87
Sydney	56	90
Canberra	61	95
Brisbane	74	122
Perth	76	110

Notes
* From Urban Water Management Plan.
** Does not include unaccounted for water (e.g., system leaks)

WATER AND DEVELOPMENT: THE PEOPLE'S REPUBLIC OF CHINA

In early stages of economic development, little investment is placed in maintaining environmental quality. "Modern" diseases such as cancer and diabetes and "epidemics" such as obesity and tobacco use generally increase alongside the decline of traditional maladies, like waterborne infectious diseases. This is the "Risk Transition" concept.

According to the World Resources Institute,[9] the Risk Transition interval is illustrated by the recent history of China. Declining but still substantial traditional illnesses and increasing modern ailments pose a major challenge to China's health-care system. The relationship between water and health will strongly influence China's present and future development.

Approximately 700 million Chinese—over half the population—must use water that contains levels of animal and human waste that fail to meet minimum drinking water quality standards. Three reasons account for China's degraded water:

1. Rapid and unregulated expansion of industry
2. Failure to invest in infrastructure to meet growing urban water needs
3. Reliance on sewage effluent to irrigate crops

Industries scattered throughout rural China are gradually replacing obsolete state-owned concerns.

These industries are unregulated and have few if any pollution-control facilities. As a result, such contaminants as excess organic matter, acids, alkalis, the nutrients nitrogen and phosphate, organic and inorganic chemicals, and heavy metals such as lead, cadmium,

[9] World Resources Institute, www.wri.org.

mercury, and chromium are among the major water pollutants detected in rural water bodies. The impacts of these toxins fall heaviest on the very young and very old.

Urban areas likewise experience heavy pollution loads. Each year, more than 30 billion tons of urban sewage is discharged into rivers, lakes, or seas with virtually none receiving even rudimentary treatment. Chinese researchers report that untreated sewage usually contains dangerous levels of pathogenic microorganisms. Rivers in cities are also polluted by inorganic and organic chemicals, many of them known to be carcinogenic, mutagenic, or both. Some researchers found associations between gastric diseases and elevated cancer rates in areas irrigated with sewage as well as industrial wastewater. When municipal water is chlorinated (to kill microbial pathogens), halomethanes such as chloroform and trihalomethane are formed by the reaction between chlorine and organic matter. These substances in high enough concentrations may pose a serious cancer risk (up to one in one hundred).

Agriculture is a major source of water pollution in China. Since the 1980s, many farmers have adopted so-called "green revolution" technologies common to western industrial agriculture: intensive fertilizer and pesticide application together with hybrid seeds. As a result, the use of pesticides and fertilizers has been increasing in China. Fertilizer use increased by at least one-third during the 1990s, but inefficient application meant that up to 70 percent of the fertilizer was wasted, leading to high levels of toxic nitrate in groundwater, which can kill infants.

Yet while water pollution is a persistent threat, more than half of Chinese cities experience severe water *shortages,* which we illustrated above with Beijing. These could also be called demand *longages.*

Question 7-11: Could water shortages and water pollution be a cause for the increase in Chinese nationals trying to gain entry into the United States and other countries? Why or why not?

Question 7-12: Are cheap imports of consumer goods from China being subsidized by insufficient investment in environmental protection in that country? What information would you like to have in order to answer the question more fully?

WATER AND GEOPOLITICS: ISRAELI/PALESTINIAN WATER CONFLICT

Since the advent of Jewish repatriation to Palestine around the turn of the twentieth century, conflicts over the region's scarce water have been growing. They continue to pose one of the thorniest problems stalling a comprehensive Middle East peace. At the center of this conflict is the disposition of groundwater underlying the West Bank region, called Judea and Samaria by the Israelis, access to which is claimed by both Israelis and Palestinians.

TABLE 7-3 ■ The Safe Yields of the Three Main Aquifers in the Middle East

Aquifer system	Number of aquifers	Annual safe yield (cubic meters × 10^6/y)
Western Basin	2	350
Northeastern Basin	2	140
Eastern Basin	6	125

Further complicating the issue is the rapid population growth in the region on both sides, on the Israeli side coming substantially from immigration.

Groundwater supplies in the region are shown in Table 7-3. Groundwater is the most important source of fresh water in the region and comes mainly from aquifers located and recharged in the West Bank. Total precipitation is estimated at 2,600 millimeters, of which only about 600 millimeters per year is infiltration.

Untapped groundwater supplies as of 1998 were estimated at 78 million cubic meters (mcm) per year.

Water Supplies in Gaza

Water supplies in the Gaza Strip are even more restricted than supplies elsewhere. Other than home cisterns to catch and store scarce rainfall, groundwater is the only source of fresh water. The present safe yield of the aquifers is estimated at 65 million cubic meters per year (mcm/y), but at least 100 mcm/y is being withdrawn. Thus, these systems are being "mined" at the rate of at least 35 mcm/y. As they are in contact with seawater, the aquifers are being poisoned by saltwater intrusion and will eventually be unusable should present trends continue. Since the population of Gaza is increasing, these trends will likely worsen. As of 1995, over 60 percent of groundwater withdrawals went to irrigate crops, so should irrigation efficiencies increase, more supply would be available to protect the system from saltwater incursion.

West Bank Groundwater

The Judean/West Bank aquifer system underlies the Palestinian West Bank, the Jordan River/Dead Sea Rift, and the narrow Israeli coastal plain. Water was one of the major items of disagreement between Israelis and Palestinians during the negotiations leading to the interim "Oslo B" agreement, signed in Washington, D.C., in 1995. While Israel recognized Palestinian water rights in the West Bank, the interim agreement left details unresolved. The CIA estimated the Palestinian population in West Bank and Gaza at 4.3 million in 2012.

Based on present water allocations, the average Palestinian has access to less than 25 cubic meters of water per year.

Question 7-13: How does this figure compare to the annual water needed to preserve human health cited earlier?

Were a goal to be set to allocate to each Palestinian the minimum of 50 cubic meters per year recognized to meet basic domestic needs, more than 70 mcm/y of supply would have to be found in a region that is already near capacity in terms of water use.

Approximately 25 percent of Israel's water supply comes from West Bank aquifers. The present allocation of groundwater is described as "the abilities of the strong to impose their wills on the weak."[10] Furthermore, Israeli surface-water use has polluted the Jordan River downstream from Israeli diversions, rendering the water unsuitable for West Bank farmers.

Question 7-14: Do the Israelis have any responsibility to provide adequate water supplies to Palestinians in Gaza and the West Bank? Explain reasons for your answer. Does your answer exacerbate tensions in the Middle East or relieve them? Why?

Future water needs cannot be met by increasing supply from presently existing sources. Thus, conservation, new supplies, or a combination of both will be needed, along with measures to address population growth, if the issue of Middle East water allocations is to be fairly settled. And if the issue is not fairly settled, it is unlikely that the region will experience a long-term decline in hostilities.

Question 7-15: Summarize the main points of this chapter.

Question 7-16: Discuss whether the water demand in the Middle East is sustainable. If it is not, what could be done to make it sustainable?

FOR FURTHER THOUGHT

Question 7-17: Do an Internet search for anecdotal illustrations of the nature of conflicts over water access and quality. How do authorities propose to address these issues?

Question 7-18: Do a similar Internet search for examples of water conflicts—both international and intranational. Summarize your findings in a paragraph or two.

[10] Environmental Degradation and the Israeli-Palestinian Conflict, http://www.arij.org/.

Question 7-19: Find examples of water waste either in the United States or abroad. You could focus on agriculture, industrial, or domestic use. Summarize in a paragraph or two what you found.

Question 7-20: Do another Internet search using the term "water pricing." Cite examples of irrational pricing or conflicts over water prices.

Question 7-21: Find out how your state or province adjudicates water conflicts or regulates "ownership" of water. Summarize in a paragraph what you found.

Question 7-22: Pick one of the waterborne diseases mentioned in this chapter and find out more about it. What causes it? How can it be prevented or treated? What is the annual cost of the disease to a country or globally in lives or money? Summarize in a paragraph or two what you found.

Question 7-23: The film Erin Brockovich, based on a book, concerns the effects of chromium in a community's water supply. Watch the film or read the book and record your responses.

Question 7-24: Watch the documentaries *Flow* (2008) and/or *Tapped* (2010), which focus on privatization of water supplies and whether access to clean water is a human right or a commodity. Use evidence from the videos as the basis to discuss the costs and benefits of privatization of water supplies, and whether you support it.

CHAPTER 8

WHACKER MADNESS?
THE PROLIFERATION OF TURFGRASS

KEY QUESTIONS

- What is the economic impact of turfgrass and lawns?
- What are the positive and negative environmental impacts of turfgrass proliferation?
- How significant is pollution caused by turf and lawn maintenance?
- Is water use in turf maintenance a significant environmental issue?
- Are millions of acres of turf consistent with sustainability?

BACKGROUND

A reader wrote to advice columnist Ann Landers:[1]

> Dear Ann:
>
> We bought a lakefront home in the woods thinking we would get away from the weekend Lawn Rangers, to no avail. Our neighbors have this golf course mentality, which is positively maddening. If it isn't the lawn tractor, it's the weed whacker, the leaf blower, the power mulcher, the lawn vac or a chain saw. . . . Enough already!
>
> We spend a minimal amount of time mowing down the weeds . . . because there are better things to do in life than mow lawns and contribute to noise pollution. They probably refer to us as—
>
> The Schlocky Neighbors in Knowlton, Wisc.

Question 8-1: Before continuing, contrast activities involved in "gardening" with those involved in "lawn maintenance."

[1] December 19, 1997.

LAND-USE CHANGES IN NORTH AMERICA

Before the arrival of Europeans, most of the eastern United States was hardwood forest. Changes in land use during the ensuing 400 years have been profound, as documented in sediment cores taken from Chesapeake Bay[2]. The first major change involved clearing of old-growth forest for agriculture, which was substantially complete by 1920, at least in the watershed of Chesapeake Bay.[3]

Land-use changes continue, most notably the "suburbanization" of much of the United States, which was a hallmark of the twentieth century. Accompanying this profound shift in land use has been the conversion of large areas of the United States to turfgrass.

We owe to the twentieth century the idea of a "smooth, green carpet as a necessary adjunct to the perfect home;" that is, a lawn.[4] Here is a sense of the importance of turfgrass today.[5]

- More than 1 million acres of farmland is devoted to grass seed and sodgrass cultivation.
- There are four million acres of managed turf in chronically water-short California
- In the Chesapeake Bay Watershed (CBW), turf comprises 5.3–9.5% of total watershed area.[6]
- In Maryland, 23% of the state's area within the CBW was turf in 2010.[7]
- Overall, there were about 128,000 square kilometers of managed turf in the United States as of 2012.[8]

Question 8-2: Calculate the acreage in hectares (ha) in managed turf in the U.S. based on 128,000 km^2. (There are 10,000 square meters/hectare).

The eastern United States was virtually cleared of virgin forest between 1850 and 1920 and has been partially reforested since 1920. Even so, less than half the forest cover of 1620 remains.

Question 8-3: In Virginia, lawn area increased from 714,000 acres in 1998 to 1,048,000 acres in 2004.[9] What percentage of Virginia (42,769 square miles, 27,372,160 acres) was turf as of 2004?

---

[2] Cooper S.R., and G.S. Brush. 1991. Long-term history of Chesapeake Bay anoxia. *Science, 254:* 992–996.

[3] See U.S. Energy Information Administration. 1996. *Emissions of greenhouse gases in the United States,* U.S. Dept. of Energy, Washington, D.C., p. 65, Figure 11. This publication includes maps showing the approximate extent of forest cover in the United States for 1620, 1850, 1920, and 1992.

[4] http://web.archive.org/web/20110125025810/http://www.edf.org//article.cfm?contentID=984.

[5] From the U.S. Department of Agriculture (USDA), World Resources Institute. 1997. *World Resources 1996–97; A Guide to the Global Environment* (New York: Oxford University Press); TPI Turf News 2005.

[6] Chesapeake Stormwater Network Tech. Bull #8, April 2010.

[7] Ibid.

[8] http://earthobservatory.nasa.gov/Features/Lawn/lawn2.php.

[9] www.vaturf.org.

Question 8-4: By 2011, 8.2% of that part of Virginia within the Chesapeake Bay Watershed (CBW) was turf. Roughly 60% of Virginia's area is within the CBW. Estimate how much of Virginia was turf in 2011.

Question 8-5: Assume the rate of increase from 1998 to 2004 to be constant. When would the entire state of Virginia be turf? Use $t = (1/r) \ln(N/N_0)$.

WHAT IS TURF?

Turf is a grass *monoculture*. Even though there are over 5,000 species of grasses worldwide, sodded lawns contain very few species, and most warm-season grasses are dominated by a single species. Turf comes from one of two sources: (1) grass seed sown on a plot of soil, or (2) pre-grown turf rolls (also called sod) produced on turf "farms," (Figure 8-1, Table 8-1) which represent a significant land use in parts of the Pacific Northwest.

FIGURE 8-1 Growing sod in the desert in Las Vegas, NV. (Courtesy of Robert Holmes/ Corbis Images.)

TABLE 8-1 ■ Willamette Valley, Oregon Turf Acreage

Year	Grass Seed Acreage	Sales
1988	332,610	$ 190 Million
1997	410,510	$ 300 Million
2008	489,600	$ 469 Million
2010	375,665	$ 228 Million

Replacing turf with another crop is not an easy task in Oregon's Willamette Valley. Grass tolerates the extremely wet winters and very dry summers, but other crops do not do as well. Moreover, installing drainage systems in the Valley's volcanic, clay-rich soils to move rainwater off soaked fields in winter, necessary for most food crops, can cost $1,000 to $5,000 an acre.

Question 8-6: What is the value in dollar sales per acre for each of the years?

Most new construction uses turf rolls from sod farms to give the house or commercial building an "instant lawn." According to the Professional Lawncare Network (PLN),[10] lawns do the following:

1. Produce oxygen as the plants photosynthesize: "625 square feet of lawn provides enough oxygen for one person for an entire day."
2. Cool the temperature: "On a block of eight average houses, front lawns have the cooling effect of 70 tons of air conditioning."
3. Control allergies by controlling dust and "replacing plants to which many people are allergic."
4. Absorb gaseous pollutants CO_2 and SO_2.
5. Trap particles (up to 12 million tonnes annually).
6. Protect water quality by filtering runoff.

Question 8-7: Evaluate PLN claim 1, using critical thinking principles. Certainly turf produces more oxygen than dirt or asphalt, but how does turf compare with forest or other land uses? What additional information do you need to assess this claim?

Question 8-8: Does claim 2 tell you what the cooling effect of turf is being compared to? Asphalt? Forest? Baked beans? Is the statement accurate? Precise? Clear? Ambiguous? What additional information do you need? Discuss.

[10] www.landcarenetwork.org.

Question 8-9: Analyze claim 3. Grass allergies are one of the most widespread and severe allergies in the United States. (See "For Further Thought" questions at the end of this chapter.)

Adverse Effects of Lawns and Turf

Although sod is clearly better at erosion control than bare dirt, much turf has replaced forest, which is much more efficient at all the attributes described by the PLN. Sod has replaced agricultural land as well (see "For Further Thought" section for details).

Most lawns receive inappropriate doses of fertilizer, insecticide, and herbicide. For example, an annual nitrogen application of 3–4 lbs/1000 ft^2 will likely be necessary for new turf established on a site devoid of organic matter and plant nutrients, like most new suburban home sites. Improper fertilization of lawns can release substantial nitrate and phosphate into streams, fueling algal growth that leads to depletion of oxygen (hypoxia or anoxia). This may cause fish kills and contribute to infestations such as the *Pfiesteria piscicida* outbreak in 1997 that caused the closure of several streams tributary to Chesapeake Bay.[11]

Increases in turf area are associated with *sprawl development* (see Chapter 13). Sprawl replaces agricultural land and woodland. Woodland is irreplaceable as wildlife habitat, is essential for aesthetics, and can help counter global warming induced by greenhouse gases.

"Caring" for turf has not-so-quietly become one of the nation's most polluting activities. Lawns require regular maintenance, according to the PLN. They should be watered when dry, mowed at regular intervals, and fertilized twice a year. Since most lawn-care devices (mowers, trimmers, blowers, weed whackers, and the like) are gas-powered, and many still have two-cycle engines, they burn fuel—a lot of it. In fact, a two-cycle gasoline-powered lawn mower run for an hour emits about the same amount of smog-forming emissions as forty new cars run for an hour, according to the State of California's Air Resources Board (ARB). Moreover, the World Health Organization (WHO) reports that the harmful effects of noise pollution are second only to polluted air in EU countries.[12]

Suburban lawns and gardens receive more pesticide applications per acre—3.2 to 9.8 pounds—than agriculture, which receives 2.7 pounds per acre on average. Of thirty commonly used lawn pesticides during the 2000s, nineteen were possible carcinogens, thirteen linked with birth defects, twenty one with reproductive effects, fifteen with neurotoxicity, twenty six with liver or kidney damage, and eleven have the potential to disrupt the endocrine (hormonal) system. The National Academy of Sciences estimates 50 percent of lifetime pesticide exposure occurs during the first five years of life.[13]

---

[11] To assess the relative environmental impact of urban/suburban, pasture, cropland, and forested land uses, go to the EPA Chesapeake Bay Program website, http://www.epa.gov/Region3/chesapeake/ and http://www.chesapeakebay.net/.

[12] *The Economist*, June 30 2012, p. 82.

[13] www.beyondpesticides.org.

The "Lawn Care" Industry

Increasing numbers of suburban households with working spouses find themselves with a decreasing interest in, or time for, lawn "maintenance." U.S. homeowners, including a growing number of retirees, are turning to lawn, landscape, and tree care professionals in record numbers. More than 29 million households regularly hired a "landscape service" by 2000. Cynics refer to it as one of the last "growth" industries in America. Lawn-care devices themselves, such as blowers, mowers, and trimmers ("weed whackers"), most of them gas-powered, represent a $5-billion-a-year growth industry. According to the Lawn Institute, Americans spent $6.4 billion caring for lawns in 2011.[14]

Is all of this cause for concern or congratulation?

Question 8-10: According to the Turf Producers International, a "well-maintained" 10,000 ft^2 lawn will generate around 1 ton (900 kg) of grass clippings annually.[15] On average, what is the weight of grass clippings (in pounds) produced per square foot of lawn per year? Restate your answer as kilograms per square meter.

Question 8-11: Based on this and the information in Questions 8-3 and 8-4, how many kilograms of grass clippings were produced from Virginia lawns in 2011?

Runoff from lawns has contributed to the widespread presence of pesticides in streams and groundwater. The chemical 2,4-D is the herbicide most frequently detected in streams and shallow groundwater from urban lawns.

Question 8-12: According to the city government of Mesa, AZ, lawns require 40–55 in. of rain or irrigation water per year, and Mesa gets around 8 in. of rain a year. Calculate the amount of water necessary to irrigate 5,000 ft^2 of lawn each year in Mesa, in addition to the rain. Express the amount in liters per day.

Question 8-13: How does this figure compare with the U.S. 2004 per capita water consumption of 1,800 gallons per day?[16] Convert your answer to liters.

---

[14] http://www.thelawninstitute.org/.

[15] www.turfgrasssod.org.

[16] U.S. Bureau of the Census, 2006. Statistical Abstract of the U.S.: 2006. www.census.gov/compendia/statab; www.waterfootprint.org/Reports/Hoekstra_and_Chapagain_2007.pdf

ALLERGIES AND INDIVIDUAL RIGHTS

Many southwestern cities report increases in residents with allergies. In Tucson, "residents have twice the national rate of respiratory allergies."[17] Experts attribute the increases to people who move west from eastern states and bring a taste for lawns and eastern trees with them.

Some cities are beginning to ban certain plant species. Tucson has banned Bermuda grass (*Cynodon*), a key species in many lawns, and many western cities discourage turf plantings while encouraging "xeriscaping," that is, planting native drought-resistant plants in lawns to cut down on water demand.

However, some residents object that their rights are being violated. When Albuquerque's city council banned several types of trees, a resident responded, "I get pretty sick of people . . . saying that this is a desert and we have to live like it's a desert. We do not have to live like it's a desert. I love trees and I choose to live near them."[18]

Question 8-14: Do you think that city governments have an obligation, or the authority, to tell residents what they can or cannot plant? Do you think that people should have a right to plant whatever they like on their own property? Defend your reasoning.

Question 8-15: Health-care costs consistently have risen faster than the rate of inflation, and part of the cost of health care is treating people with severe allergies. Do you feel that those who insist on their right to plant whatever they choose on their property should be required to pay for the added burden on the health-care system that their actions cause or contribute to? How? Explain your reasons. Is this an environmental issue? Why or why not?

GAS POWERED LEAF BLOWERS

"Leaf blowers: the most annoying . . . power tool ever invented."[19]

The gas-powered leaf blower was introduced to the United States as a lawn maintenance tool, but it gained little acceptance until drought conditions in California during the late 1970s led to restrictions on outdoor water use. In 2011, sales of gasoline-powered and electric leaf blowers exceeded 4.9 million units, at an estimated cost of $535 million.[20] A typical gas-powered blower uses about a third of a gallon of gasoline per hour of operation.

---

17 Mendoza, M. Southwestern communities find greening desert is something to sneeze at. *Washington Post*, November 10, 1996.

18 Ibid.

19 www.noisefree.org.

20 http://www.homechannelnews.com/article/leaf-blowers-numbers.

By 2011, more than one hundred cities in California alone had banned or restricted leaf blowers. The California Air Resources Board advises citizens to avoid leaf blowers.[21]

Advantages and environmental disadvantages of leaf blowers are summarized below.

Emissions from Leaf Blowers

Two-stroke engines power most gas leaf blowers. The two-stroke engine has several advantageous attributes. They are lightweight in comparison to the power they generate and operate in any position. Multipositional operation is made possible by mixing the lubricating oil with the fuel, which in turn leads to a major disadvantage: high exhaust emissions. As much as 30 percent of the fuel/oil mixture is exhausted unburned. Thus, exhaust emissions consist of both unburned fuel and products of incomplete combustion. The major pollutants from a two-stroke engine are oil-based particulates, a variety of hydrocarbons, and carbon monoxide.

Despite government-required reductions in emissions, in 2011, edmonds.com reported that consumer-grade leaf blowers still emit as much as 23 times the toxic carbon monoxide of a Ford F-150 truck. A two-stroke engine produces relatively little NOx.[22]

Some of these emissions are *toxic air contaminants,* as defined by the EPA. They include benzene, 1,3-butadiene, acetaldehyde, and formaldehyde.

Table 8-2 summarizes the ARB's best estimate of pollution from lawn-care devices, and, based on regulations effective in 2000, the contributions of leaf blowers to the state's air pollution in 2010. The ARB concluded, "lawn and landscape contractors, homeowners using a leaf blower, and those in the immediate vicinity of a leaf blower during and shortly after operation, are exposed to potentially high exhaust, fugitive dust, and noise emissions from leaf blowers on a routine basis." And, "the day-in-day-out exposure to [significant levels of] PM10 could result in serious, chronic health consequences in the long term." In 2009, ARB concluded that exposure to fine particulates, produced by diesel engines and devices like leaf blowers, were responsible for the "premature death" of 7,000–11,000 persons per year.[23]

Question 8-16: Should the health care "industry" in this country have to bear the burden of the effects of air pollution from leaf blowers on operators and bystanders? Justify your answer.

TABLE 8-2 ■ Inventory of Leaf Blower Exhaust Emissions (tons per day)

Assessment of Two-stroke Engine Pollution	Leaf Blowers 2000	Leaf Blowers 2010	All Lawn & Garden, 2000
Hydrocarbons, Reactive	7.1	4.2	50.24
Carbon Monoxide (CO)	16.6	9.8	434.99
Particulates (PM10)	0.2	0.02	1.05

(*Source:* California ARB)

21 www.arb.ca.gov/msprog/leafblow/leafblow.htm.
22 http://www.arb.ca.gov/html/brochure/pm10.htm.
23 Ibid.

Question 8-17: Is it a proper role of government to monitor devices produced by corporations and to set standards for their operation, or should manufacturers be required to demonstrate that their products will cause no harm to the environment before they are allowed to produce and distribute these devices? Explain your answer.

Effects of Resuspended Dust

Two additional sources of emissions from leaf blowers are "fugitive" (resuspended) dust and noise. We will consider dust next.

Leaf blowers are designed to move relatively large materials at hurricane-force wind speeds, and hence will also move much smaller particles, including those below 30 micrometers in diameter, which are not visible to the naked eye.

A Sebastopol CA environmental group reported that a leaf blower operated for one hour could suspend 5 pounds of dust, some of which could take days to resettle.[24]

An Orange County, California grand jury found that "fecal material, fertilizers, fungal spores, pesticides, herbicides, pollen, and other biological substances" were found in the dust resuspended by leaf blower use.[25] The ARB estimated that streets and sidewalks could contain about 3 grams of fine sediment per square meter, which would be blown into the air by a leaf blower. In addition, the chemical analysis of paved road dust showed small percentages of toxic metals arsenic, chromium, lead, and mercury, along with particles from tire and brake wear. Fine latex particles from tires are a known human allergen, but the effect of resuspending the material is unknown. Table 8-3 shows air pollution from leaf blowers compared to motor vehicles. Leaf blower users often blow dust and debris into the streets, leaving the dust to be resuspended by passing vehicles.

Question 8-18: The parking lots of the Potomac Mills Mall, Prince William County, Virginia, measure approximately 270,000 square meters. Based on an average value of 3 grams per square meter of dust, how much dust could be resuspended by "cleaning" the parking lots using leaf blowers daily? Report your answer in kilograms.

TABLE 8-3 ■ Air Pollution from Leaf Blowers Compared to Motor Vehicles (ARB report)

	Leaf Blower Exhaust Emissions, g/hr	**Exhaust Emissions New Light-Duty Vehicle,***
Hydrocarbons	199.26	0.39
Carbon Monoxide	423.53	15.97
Particulate Matter	6.43	0.13
Fugitive Dust	48.6–1031	N/A

* New light-duty vehicle represented vehicles one year old, 1999 or 2000 model year, driven for one hour at 30 mph.

[24] www.progressivesource.org.

[25] www.arb.ca.gov/msprog/mailouts/msc0005/msc0005.doc.

NOISE POLLUTION

We briefly referred to this earlier. In addition to damaging hearing, noise may cause other adverse health impacts. These include sleep disturbance, changes in performance and behavior, and acute annoyance. In fact, leaf blower noise may have contributed to at least one murder. According to the Bergen (NJ) *Record*, in May 2000 "a woman killed her 74-year-old neighbor by repeatedly running him over with a car, after their latest dispute, which involved his use of a leaf blower, police said."[26] The victim had complained to a neighbor earlier that the alleged assailant, wearing a dust mask, had threatened him earlier with a pitchfork due to the noise, emissions, and dust from the victim's leaf blower.

Long-duration, high-intensity sounds are the most damaging and usually perceived as the most annoying of all sounds. High-frequency sounds, up to the limit of hearing, tend to be more annoying and potentially more hazardous than low-frequency sounds.

Figure 8-2 shows some common noise levels and reference values from leaf blowers.

Based on extrapolations of EPA data, at least 3 million people nationwide could be routinely exposed to leaf blower noise at annoying levels. For California, the figure could exceed 300,000, based on the urban/rural population ratio and scaling of population values.

Anti-Noise Legislation

The Federal Noise Control Act of 1972 sought "to promote an environment for all Americans free from noise that jeopardizes their public health and welfare." The EPA is charged with implementing this law.

About 13 percent of Californians live in cities that ban the use of leaf blowers, and six of the ten largest California cities, including Los Angeles, have ordinances that restrict

Perceived Sound Level	Sound Level		Examples	Leaf Blower Reference
	dB	μPa		
PAINFULLY LOUD	160	2×10^9	fireworks at 3 feet	
	150		jet at takeoff	
UNCOMFORTABLY LOUD	140	2×10^8	threshold of pain	Occupational Health and Safety Administration limit for impulse noise
	130		power drill	
	120	2×10^7	thunder	
VERY LOUD	110		auto horn at 1 meter	90–100 dB leaf blower at operator's ear
	100	2×10^6	snowmobile	
MODERATELY LOUD	90		diesel truck, food blender	90 dB Occupational Health and Safety Administration permissible exposure limit
	80	2×10^5	garbage disposal	
	70		vaccum cleaner	82–75 dB leaf blower at 50 feet
QUIET	60	2×10^4	ordinary conversation	
	50		average home	
VERY QUIET	40	2×10^3	library	
	30		quiet conversation	
	20	2×10^2	soft whisper	
BARELY AUDIBLE	10		rustling leaves	
	0	2×10^1	threshold of hearing	dB = decibels μPa = micro Pascals

FIGURE 8-2 Noise levels from common sources and some leaf blower comparisons. (Source: California ARB)

[26] www.NorthJersey.com.

or ban leaf blowers. All together, to summarize, about one hundred California cities have ordinances that restrict either leaf blowers specifically or all gardening equipment generally, including cities with bans on leaf blower use.

Question 8-19: Approximately 14 million new motor vehicles are sold in the United States each year, each of which contains a catalytic converter that costs approximately $300. Calculate the total amount Americans pay for catalytic converters to take pollution out of the air, and then comment on whether it makes economic sense to put that pollution back in the air in the form of toxics from leaf blowers. Defend your answer.

Question 8-20: Summarize the positive and negative attributes of millions of acres of American lawns. Be sure to justify or include evidence for your choices. In what ways might lawns be compatible with the principles of sustainable communities?

Question 8-21: Summarize the main points of this chapter.

For Further Thought

Question 8-22: Go to www.noisefree.org, and read representative stories on leaf blowers. Summarize in a couple of paragraphs what you found.

Question 8-23: Contact your city or county department of waste management. Do they accept lawn debris or waste? Do you think this is an appropriate use of public funds?

Question 8-24: Go to EPA's website www.epa.gov. Find out how serious emissions from gasoline-powered lawn-care devices are, and what regulations are in effect or proposed to control emissions from gasoline-powered lawn-care devices.

Question 8-25: Contrast the impact of gasoline-powered with electric lawn-care devices. What advantages do electric devices provide? What disadvantages?

FIGURE 8-3 The Reserve Golf Course in Pawleys Island, SC. This course was designed to be environmentally sensitive and has much less turf than a typical golf course. (Courtesy of The Reserve Golf Club, McConnell Golf, LLC.)

Question 8-26: Golf courses typically replace fields or forests with turf. Some newer courses (Figure 8-3) are reducing the area of turf. Research this issue. Check the golf courses in your area. Are they constructed and maintained in a sustainable way?

CHAPTER 9

OIL AND NATURAL GAS

KEY QUESTIONS

- What are proven reserves of oil and gas?
- How fast are we consuming oil and gas?
- Where are global reserves of oil and gas?
- What is hydrofracturing, and why is it controversial?
- What are the environmental effects of oil and gas use?
- What are the issues surrounding drilling for oil in the ANWR?
- What is *peak oil,* and has it been reached yet?
- How is oil and gas use related to sustainability?

BACKGROUND

Oil is the energy basis of modern industrial society. For over fifty years after the first successful well was drilled in the mid-nineteenth century, oil use was insignificant. Although he thought his vehicles would be powered by peanut oil, Henry Ford changed all that with his mass-produced automobiles. World War I generated immense demand for gasoline-powered vehicles. During the period 1919 to 1949, oil gradually overtook coal as the most important fuel source in the United States. Today, oil provides 40% and natural gas provides more than 23% of U.S. energy. Oil provides virtually all of our transportation fuel.

By 2012, global oil demand was around 88 million barrels (1 barrel = 42 U.S. gal) per day, with the United States responsible for more than 18 million barrels per day.[1]

Question 9-1: What percent of global oil demand is accounted for by the United States?

Question 9-2: What percent of the world's 7.05 billion population is represented by the 310 million of the U.S.?

[1] U.S. Energy Information Administration, www.eia.gov.

Because of growing demand for oil from Asian economies, especially India and China, as well of political turmoil in oil producing countries, mainly the Middle East, oil prices have risen sharply. Chinese demand alone accounted for 9.4 million barrels a day (mb/d), more than half that of the U.S. As recently as 1994, Chinese demand was only around 3 mb/d. Indeed, since 2008, demand in Brazil, Russia, China, and India has increased by 3.7 mb/d, while demand in the U.S. and Eurozone countries has fallen by only 1.5 mb/d.[2]

ORIGIN, DISTRIBUTION, AND EXTRACTION OF OIL

Oil is not a renewable resource. The oil fields of today originated tens of millions of years ago when organic remains were buried within sediments in the absence of oxygen. The organic matter was subjected to a critical combination of pressure (caused by deep burial) and increased temperatures. Over millions of years, the organic matter reorganized into more volatile organic molecules that we call *crude oil* or *petroleum,* usually accompanied by *natural gas.* To be extracted and used, petroleum must migrate upward, out of its deeply buried *source bed,* and into rock strata or a favorable geological structure that will trap the oil and prevent its escape (Figure 9-1).

Sometimes, pressure forces the oil all the way to the surface, where it forms *seeps,* which were known to Native Americans. The nineteenth century's oil discoveries came when "wildcatters" (oilmen who drilled speculative wells) simply drilled holes into the rocks underlying oil seeps.

Once extracted from underground reservoirs, oil is processed to remove other fluids, such as water, hydrogen sulfide, and natural gas, and then it is sent to refineries. There the oil is heated in the absence of oxygen to break or "crack" the molecules into lighter forms, which emerge as products such as gasoline, diesel fuel, or asphalt. Much, but not all, associated sulfur is usually removed at refineries as well. The sulfur that is left in motor fuels when burned forms oxides of sulfur (SOx), a toxic air contaminant, which cannot be removed by the present generation of catalytic converters. The oil may then be shipped by pipeline or oil tanker to destinations around the world.

HOW MUCH OIL IS LEFT?

Calculating the amount of oil present in all known world oil fields that can be extracted at a profit using present technology yields a value called *proven oil reserves* (Table 9-1).

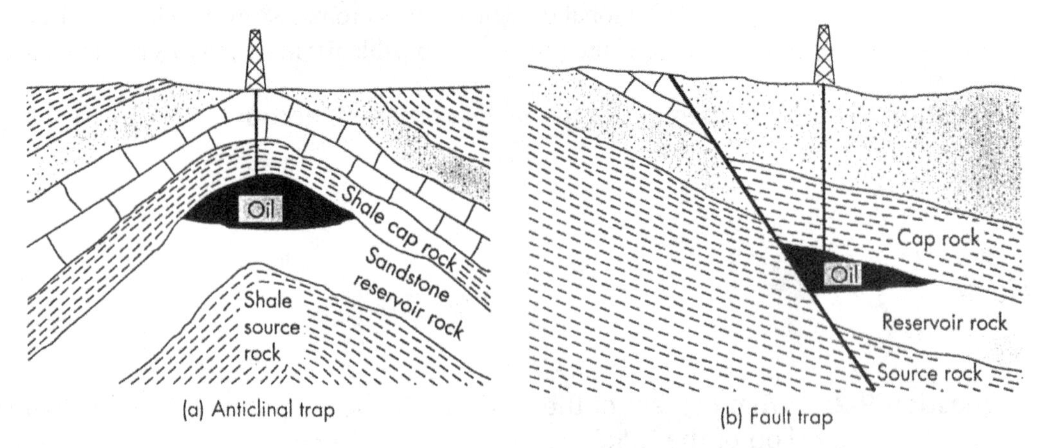

FIGURE 9-1 Typical oil and gas "traps." Oil, being lighter than water, floats to the top of the reservoirs. The oil is usually associated with natural gas as well. (Keller, E.A. 2000. *Environmental Geology* 8th ed., Fig. 15.10, page 411. Prentice Hall, Upper Saddle River, NJ. Courtesy of E. A. Keller/Pearson Education.)

[2] *The Economist*, 6/23/12 p. 73.

TABLE 9-1 ■ Proven Oil Reserves (billion barrels) by Country, 2011[1]

1. Saudi Arabia	262
2. Venezuela	211
3. Canada*	175
4. Iran	137
5. Iraq	115
6. Kuwait	104
7. U.A.E.	98
8. Russia	60
9. Libya	46
10. Nigeria	37
11. Kazakhstan	30
12. Qatar	25
13. U.S.A.	20

*Canada has over 130 billion barrels of waxy petroleum solids in sedimentary deposits called "tar sands." Although they could be, and are, extensively mined, they are not conventional deposits.

[1]Source: US CIA.

Organizations such as the U.S. Department of Energy and the International Energy Agency collect and publish such information. World proven oil reserves are less than the actual oil in place. Variable amounts of oil in underground reservoirs always remain in the rock. However, new technology may increase oil recovery, or the price may go up making it economic to spend more money to get more oil out. Moreover, companies frequently underestimate the amount of oil in a field or may not publicize the actual amount for competitive reasons. For example, in 1970 BP estimated its Forties North Sea oil field contained 1.8 billion barrels of proven reserves. However, by 2012, BP, and new owner Apache, had produced 4.0 billion barrels from the field!

If oil price rises, marginal fields—those not economic to develop at current prices—may become profitable and may be added to reserves. Chevron Oil Corporation estimated in 1991 that 6,700 billion barrels of oil could be ultimately recovered, assuming *a price of $60 per barrel in 1990 dollars.*[3]

Question 9-1: According to the Department of Labor, the consumer price index (CPI) is a measure of the representative cost of a "basket" of goods and services. It stood at 127.4 in 1990, and had reached 230 by mid-2012.[4] Therefore, by what percent had the CPI increased between 1990 and 2012?

[3] Holyoak, A.R. Written communication.

[4] U.S. Department of Labor Bureau of Labor Statistics (http://www.bls.gov/bls/.

Question 9-2: To determine the 2012 price that would be equivalent to a 1990 price of $60 per barrel, first multiply the 1990 minimum price by the percentage you just calculated. What would be the 2012 price needed to "guarantee" 6,700 million barrels of ultimately recoverable oil?

Question 9-3: For a world price of $100+ per barrel to stimulate increased oil production, the price would have to be maintained long enough for companies to be able to make investment decisions on it, since producing oil from a new discovery can take ten years. What do you think would happen to oil exploration and ultimately production if the world price were to fall precipitously?

TECHNOLOGY ENHANCES OIL DISCOVERIES AND PRODUCTION

Over the past several decades, technology has helped discover new oil and gas fields. It has also increased ultimate recovery of the oil and gas in place, both in known fields and new discoveries. Companies use seismic data and powerful computers to create 3-D images of deeply buried geologic structures more than 5 kilometers below the surface. And recent advances in drilling techniques have lowered production costs and increased oil and gas yields from known fields. Using new techniques, companies have increased the ratio of discoveries to "dry holes." Dry holes are exploratory wells that didn't hit oil and can cost up to $15 million each!

For the past twenty-five years, with only two exceptions, estimates of ultimately recoverable oil have ranged from around 1,800 billion barrels to around 2,400 billion barrels. We will discuss a new method of natural gas exploration and production, hydrofracturing, below.

Where Is the World's Oil?

As you can see from Table 9-1, the Persian Gulf region contains more than half of the world's proven oil reserves. It similarly produces half the world's oil. The Gulf region is such an extraordinary source of oil because it has huge oil *fields* and extremely low costs of exploration and production. For example, it can cost as little as a few dollars per 42-gallon barrel in the Persian Gulf to produce oil. In new oil fields of North America and the North Sea of Europe, costs can range from $20 to $60 a barrel.

Question 9-4: What is today's world price for oil? Check the web, a daily newspaper, or a business network.

We can determine how long world proven reserves (1,300 billion barrels at the end of 2011) will last by simply dividing the total oil by the annual production—approximately 32 billion barrels per year.[5] Note that when we do this, the resultant number is years since "barrels" cancel out.

[5] U.S. Central Intelligence Agency (CIA), www.cia.gov.

This number represents the total number of years from 2011 that oil could be produced at current rates before it is all gone, *assuming* constant production *and* demand. In reality, the depletion of a resource like oil does not follow such a simple pattern. Rather, oil production will gradually decline over a long period of time.

Question 9-5: Assuming no new discoveries, in what year, starting from 2011, would the world run out of oil?

However, the rate of consumption is increasing each year, and new discoveries are being made (Table 9-2).

Consult Table 9-2. Describe the change in global demand from 1986 to 2012.

TABLE 9-2 ■ World Oil Demand[2]

Year	Consumption (million barrels per day)
1986	61.8
1987	63
1988	64.8
1989	65.9
1990	66
1991	66.6
1992	66.7
1993	67
1994	68
1995	70
1996	71.7
1997	73.7
1998	73.6
1999	74.7
2000	75.8
2001	76.4
2002	77.8
2003	79.3
2004	82.6
2005	83.7
2006	85.1
2007	85.8
2008	85.4
2009	85.6
2010	87.0
2011	88.0
2012 (est)	88.7

[2] U.S. Energy Information Administration. International oil data for crude oil production, available at http://www.eia.doe.gov.

Question 9-6: What was the average annual growth rate in oil consumption from 1986 to 2012? (The most accurate way to calculate the average rate of increase is to use the formula $r = (1/t)\ln(N/N_0)$; see "Using Math in Environmental Issues," pages 6–8).

$r = (1/27)\ln(88.7/61.8) =$

It is important to note how a tiny rate of increase can lead to such a change in oil consumption over a long period of time. The concept of doubling time, $t = 70/r$ introduced earlier, can be applied to illustrate growth of consumption.

Question 9-7: What is the doubling time for the average increase in demand you just calculated?

Question 9-8: To estimate the total amount of oil that was consumed over a given period, first convert the beginning and ending rates of consumption (say, 80 million barrels a day) to barrels per year. Then add the beginning and ending consumption rates together and divide by 2. Finally, multiply this number by the number of years. How much oil was consumed between 1986 and 2011?

Question 9-9: EPA reports that 0.43 tonnes of CO_2 is emitted per barrel of oil burned. How much CO_2 was emitted by burning this much oil between 1986 and 2011?

Author and environmental provocateur Garrett Hardin, in addressing the issue of energy supply, said, "Whenever it is thought to be impossible to limit the growth of either population or desire, *it is impossible to solve a shortage by increasing the supply*" [emphasis his].[6]

Question 9-10: Explain why you agree or disagree with this statement. Can we ever accommodate the world's increasing demand for oil? (If you have some background in economics, cite reasons why an economist might disagree with Hardin's quotation).

[6] Hardin, G. November/December 1996. Letter to Editor. *Worldwatch Magazine,*

ANWR's Oil Reserves

The American Petroleum Institute, among others, would like to open the Arctic National Wildlife Refuge (ANWR) on Alaska's North Slope to oil exploration.[7] The U.S. Geological Survey (USGS) estimates between 5.7 and 16 billion barrels ultimately might be produced over several decades.[8] Environmental scientists are concerned about (1) the impact of oil exploration on the tundra environment and (2) the impact of oil exploration and production on caribou and other migrating animals.

Question 9-11: The mean estimate by the USGS of recoverable oil in ANWR is about 10.3 billion barrels without regard to price. Based on the mid-2012 U.S. annual consumption rate of about 7 billion barrels of oil per year, how many months would these new fields last?

Question 9-12: Did we phrase Question 9-12 fairly and accurately? Why or why not? Cite evidence for your position.

Impacts of Oil Refining

Oil refineries are one of the top sources of industrial air pollution in the United States, and a dangerous place to work. For example, in 2010 four workers were killed and three severely injured in an explosion at a refinery in Washington State. And in 2005 a fire and explosion at a Texas refinery killed 15 workers and injured more than 170. Moreover, refineries are one of the largest stationary sources of volatile organic compounds, the primary component of urban smog. They are the fourth largest industrial source of toxic emissions and the single largest source of benzene emissions, which are carcinogenic.

Question 9-13: Research this issue and explain why oil refineries are so dangerous.

---

[7] American Petroleum Institute (www.api.org).

[8] U.S. Geological Survey, www.usgs.gov.

Question 9-14: Most politicians express horror at the prospect of higher oil prices. Identify advantages and disadvantages of higher prices for oil and refined products like gasoline and diesel fuel.

China became a net oil importer in 1995, and the Chinese demand for petroleum has increased significantly since 2001. Chinese oil demand was 9 million barrels a day in 2010 and was forecast by U.S. EIA to increase to 19 million barrels a day by 2020.

Question 9-15: What are the implications for the world oil price if Chinese demand sharply increases?

NATURAL GAS

Methane, CH_4, is the main constituent—about 75 percent—of natural gas, but natural gas usually also contains ethane (C_2H_6), propane (C_3H_8), and butane (C_4H_{10}). It is one of the planet's most important commodities. It presently meets more than 20 percent of world energy needs, and nearly a quarter of the energy needs of the United States. Since burning gas produces no SOx pollution, no ash or toxic emissions like heavy metals, and half the CO_2 of coal, natural gas is poised to be the fastest-growing fossil fuel source in the twenty-first century.

Natural gas can be produced in any geologic environment in which organic matter is decomposed by microorganisms in the absence of oxygen, with or without heat and pressure.

Deposits of Natural Gas

Natural gas deposits fall into four categories:

1. Natural gas is usually found with petroleum in reservoir rocks.
2. Important reserves of gas are also found in rocks with little or no oil, most recently in black shales.
3. Natural gas originates with coal and is typically found with coal deposits.
4. Methane can be formed by the action of certain bacteria in oxygen-free environments, such as waterlogged soils in permafrost regions and in deep marine sediment.

Methane deposits in deep marine sediments are called *methane hydrates*. Recent research on methane hydrates in the world's oceans point to hydrates as a potential new category of natural gas resources, though none is produced at present.

Transporting Natural Gas

Natural gas may be transported in two ways: in pipelines (the Chinese built the first ones of bamboo in the sixth century BCE) as a pressurized gas, or as a super-cooled liquid (LNG, liquified natural gas) using specially constructed tankers.

Global Gas Reserves

The U.S. EIA estimated global proven reserves of natural gas at 6,500 trillion cubic feet (TCF) in 2012. In 2011, the world used more than 113 TCF of gas, and this quantity was projected to increase at 1.6% per year through 2035. In the United States, natural gas consumption averaged around 22 TCF during 2005–2011. Reserves of natural gas are shown in Table 9-3. This does not include coalbed methane. The volume of gas in coal seams is more uncertain. What is certain is that methane seeping from coal mines is a major greenhouse gas, as well as a valuable potential resource. The presence of natural gas along with coal poses the biggest hazard to miners working underground, since it is under pressure, is odorless and colorless, and easily ignited. The Chinese government acknowledges that several thousand coal miners die yearly, and labor activists argue the figure is much higher. Most of these deaths were caused by methane explosions.

Even though burning methane produces CO_2, the primary greenhouse gas, methane is a far more powerful greenhouse gas than CO_2. Extraction of methane in deep marine sediments is not presently economically viable, although the amount of methane in such deposits is enormous.

TABLE 9-3 ■ World Natural Gas Reserves by Country 2005 and 2011 (trillion cubic feet)

Country	Reserves, 2005	Reserves, 2011
World	6,040	6,500
Russia	1,680	1,680
Iran	940	1,050
Qatar	910	900
Saudi Arabia	235	280
United States	189	277
Turkmenistan	74	270
United Arab Emirates	212	230
Nigeria	176	190
Venezuela	151	180
Algeria	161	160
Iraq	110	115
Australia	—	110
Indonesia	90	105
Kazakhstan	66	85
Malaysia	29	85
Egypt	57	75
Norway	75	72
Uzbekistan	71	65
Netherlands	65	—
Canada	62	60
Ukraine	40	40

Source: CIA World Factbook (https://www.cia.gov/library/publications/the-world-factbook/rankorder/2179rank.html)

Hydraulic Fracturing

Hydraulic fracturing ("fracking") permits gas production from rocks containing gas but with very low permeabilities, such that the gas cannot be extracted using conventional methods. Fracking injects large volumes of fluids and "proppants"(small particles of solids) at high pressure, fracturing the rock, which allows gas to escape. The fluids inject the proppants into the fractures to keep them from sealing shut when production begins. The fracking fluid is typically water-based and contains such chemicals as bactericides, buffers, fluid-loss additives, and surfactants (essentially, detergents), to make the fracturing efficient and prevent damage to the formation.

There are two problems with fracking: first, companies usually, for proprietary reasons, are reluctant to disclose the precise content of the fracking fluid, and second, large amounts of water are used, whose content is modified by mixing with water from deeply buried rocks, as well as the fracking fluid itself. Potential environmental impacts include contaminated wells, and local water supplies may be threatened. One of the nation's greatest reservoirs of "tight" shale gas, the Marcellus Shale, underlies much of the eastern United States, but production from the formation may be severely restricted if means to prevent water contamination are not perfected.[9] Indeed, half the new natural gas wells in the U.S. involve fracking. Globally, EIA estimates shale ("tight") gas reserves to exceed 60 trillion cubic feet.

Question 9-16: Global gas consumption is projected by the U.S. Energy Information Administration to grow at 1.6% percent per year. Use the doubling time formula to project how long at this rate it will take for consumption of natural gas to double from 2011's 113 TCF per year.

Question 9-17: From the perspective of global sustainability, what role should natural gas play in the world's energy future? Why?

Question 9-18: Summarize the important points of this chapter.

FOR FURTHER THOUGHT

Natural Gas from Prudhoe Bay

The Prudhoe Bay oil field in Alaska's North Slope has large reserves of natural gas, estimated by field operator BP at 35 TCF, but no way to get the gas to market. Gas cannot be transported using the existing oil pipeline, which cost over $7 billion to build in the 1970s.

[9] West Virginia Rivers Coalition www.wvrivers.org/articles/Marcellus%20Report%202010.pdf.

Question 9-19: By what two ways could the Prudhoe Bay gas be transported to a potential market?

Question 9-20: As shale gas has become more common since 2007, the price of natural gas has fallen by two thirds. How would this affect the economics of a Prudhoe Bay gas pipeline?

Question 9-21: Do you think most Americans believe that our oil and gas reserves are being depleted at a rate that will exhaust them in the twenty-first century?

Question 9-22: Ecological economists contend that were externalities to be factored into the cost of fossil energy, the price would increase substantially. Determine what these externalities might be. You could consult an article by Hubbard [10] and a paper by Roodman.[11] Also read a CEERT report.[12] A web search for "energy externalities" will yield much useful information as well. List the sources of these hidden costs, and discuss whether consumers should pay them, and, if so, what effect they might have on energy use.

Question 9-23: Consider the issue of "peak oil." The production history of a conventional oil field displays an inverted "U" shape, as shown in Figure 9-2. If one extrapolates total global production from the sum of individual fields, one arrives at a concept called peak oil. In other words, at some time in the history of global oil production, production will "peak" and then begin to decline. U.S. domestic production "peaked" in around 1970 and has been in decline since. Research the concept of peak oil for global oil reserves. When is global production forecast to peak? Assuming the concept of peak oil is accurate, discuss the extent to which oil and gas will contribute to sustainable societies.

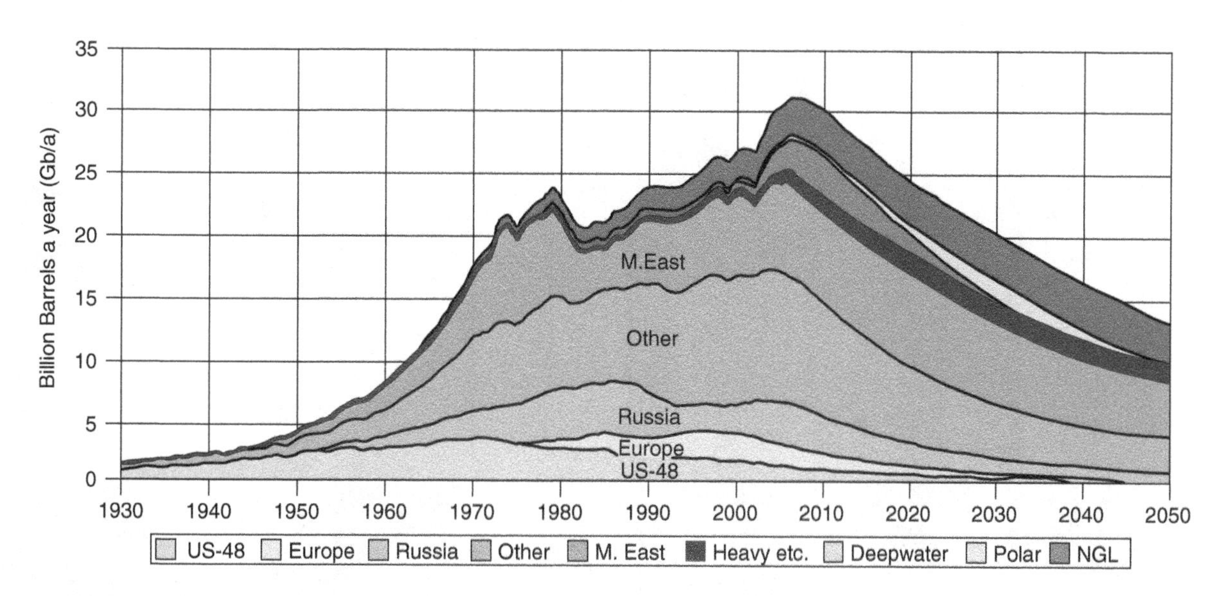

FIGURE 9-2 Hubbert's Curve, named after M. King Hubbert, whose model depicts the shape of the oil production curve.

[10] Hubbard, H.M. April 1991. The real cost of energy. *Scientific American*, 264: 36–42.
[11] Roodman, D.M. 1996. Paying the piper: Subsidies, politics, and the Environment. Worldwatch Institute paper 133. Worldwatch Institute, Washington, DC.
[12] http://www.ceert.org.

Question 9-24: Many politicians from corn-growing states like Iowa have asserted that ethanol use can free the United States from dependence on imported oil. Research the pros and cons of ethanol.

Question 9-25: Research the "Keystone XL Pipeline" and "tar sands" (also known as *oil sands* and *bitumen* sands) and summarize the main issues surrounding the controversy. Explain what Dr. James Hansen, head of NASA's Goddard Institute for Space Studies and the U.S. government's chief climate scientist, meant when he stated this about the proposed Keystone XL pipeline: ". . . if the tar sands are thrown into the mix, it is essentially game over . . ." (for the climate).[13]

Question 9-26: Dr. Hansen also wrote this about tar sands[14]: "The environmental impacts of tar sands development include: irreversible effects on biodiversity and the natural environment, reduced water quality, destruction of fragile pristine Boreal Forest and associated wetlands, aquatic and watershed mismanagement, habitat fragmentation, habitat loss, disruption to life cycles of endemic wildlife particularly bird and Caribou migration, fish deformities and negative impacts on the human health in downstream communities." In light of what you have learned about the proposed Keystone XL pipeline and tar sands, explain whether you would or would not support construction of the pipeline and further development of Canadian tar sands.

---

[13] Available at: http://www.countercurrents.org/hansen040611.htm.

[14] Ibid.

CHAPTER 10

SUSTAINABLE ENERGY:
IS THE ANSWER BLOWING IN THE WIND?

KEY QUESTIONS

- What are the main sustainable sources of energy?
- What is their potential and actual role in U.S. and global energy production?
- How is Hawaii's energy situation unique?
- Can sustainable societies be supported entirely by renewable energy?

INTRODUCTION

Renewable energy technologies presently include those that employ wind (Figure 10-1), solar, geothermal, biomass, and hydroelectric dams to produce power. But are all of these sustainable? Dams are not a truly renewable electricity source. Sediment buildup behind the dams will eventually fill the reservoir. This gives them an effective useful life of several decades to one or more centuries. And building dams has consequences: trapped sediment behind dams on the Mississippi and its tributaries is the main reason the Louisiana coast is rapidly receding, for example.

Moreover, hydroelectric dams usually have harmful effects on rivers by preventing the migration of fish species. And the construction of dams often displaces people: as many as 2 million in the case of China's Three Gorges Dam.

FIGURE 10-1 Wind turbines at Searsburg, VT (Source: U.S. EPA)

ASPECTS, ADVANTAGES, AND DISADVANTAGES OF RENEWABLE ENERGY

Renewable energy sources have the following in common:

- They are replenished by natural processes.
- They cause no direct air or water pollution since no combustion is involved.
- They require no shipment of fuels nor offshore oil exploration, so environmental disasters such as the 1989 Exxon Valdez oil tanker spill in Alaska's Prince William Sound, or the 2010 Deepwater Horizon well blowout, will not occur.
- They require no storage or use of toxic materials, like radioactive fuel, so they would make a poor target for terrorist attacks.

Renewable energy sources have, however, some significant limitations:

- Using present technologies, all must produce electricity instead of a more "portable" fuel such as gasoline or diesel. And batteries or other storage technologies that would allow renewable energy to be used extensively in the transportation sector have not yet been perfected.
- Solar power depends on sunlight. Lacking a viable means to store the electricity produced during the day, there must be a supplemental energy source at night.
- Both a minimum and maximum wind speed is required for wind power. Wind turbines are usually shut down for safety reasons if the wind speed exceeds 56 miles per hour (90 km/h). Large-scale wind power generation is limited to sites where threshold persistent winds occur.

Despite these limitations, energy production from renewable sources has increased substantially in recent years. For example, in 1999, for the first time, the world installed more new wind-generating capacity than nuclear capacity. By 2012, the Pacific Northwest had so much installed wind capacity that the Bonneville Power Authority had to occasionally mandate shutting down wind turbines.

ENERGY CONSUMPTION

Energy consumption can be measured in British thermal units (BTU), typically as units of 1 quadrillion BTU (1 quad). One BTU is approximately equal to the amount of heat necessary to raise the temperature of 454 grams (1 lb) of water by 1°F. One quad equals 10^{15} BTU. U.S. energy consumption increased from 66.4 to 80.2 quad BTU from 1970 to 1988 but declined from 100.2 quads in 2004 to 96 in 2009, before rising to 98 quads in 2010.[1]

Question 10-1: What was the percentage increase in energy consumption from 1970 to 1988?

Question 10-2: What was the average yearly increase in energy consumption over the period 1970–2010? (Use the formula $r = (1/t)\ln(N/N_0)$; see "Using Math in Environmental Issues," pages 6–8).

[1] www.eia.doe.gov/.

Question 10-3: What is the doubling time of energy consumption, based on your answer to Question 10-2? (Use the formula $t = 70/r$).

Question 10-4: Based on the 1970–2010 rate of increase, estimate U.S. energy use in 2020, and 2030. Use the compound interest formula future value = present value $\times (e)^{rt}$. What caveats would you advise about your conclusions?

Question 10-5: Total electric power generation from nonhydro renewables from 1990 to 2011 is given in Figure 10-2.[2] Assess the contribution of renewables to electricity production. What are the top three sources?

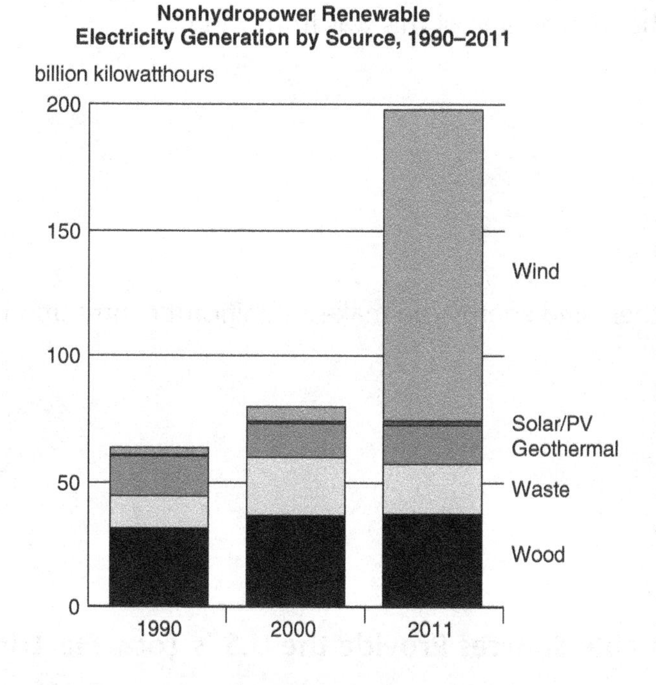

FIGURE 10-2 Nonhydroelectric renewable electricity generation, 1990–2011. (Source: U.S. Energy Information Administration, *Annual Energy Review,* Table 8.2 (October 2011), *Electric Power Monthly* (March 2012) preliminary 2011 data.)

[2] http://www.eia.gov/energy_in_brief/images/charts/nonhydro_renew_elec_large.jpg.

Question 10-6: Study Figure 10-3. Assess the contribution of renewables to the total U.S. production of electricity. What are the top four sources of U.S. electricity? Do you see cause for optimism that renewables can make a major contribution? Explain.

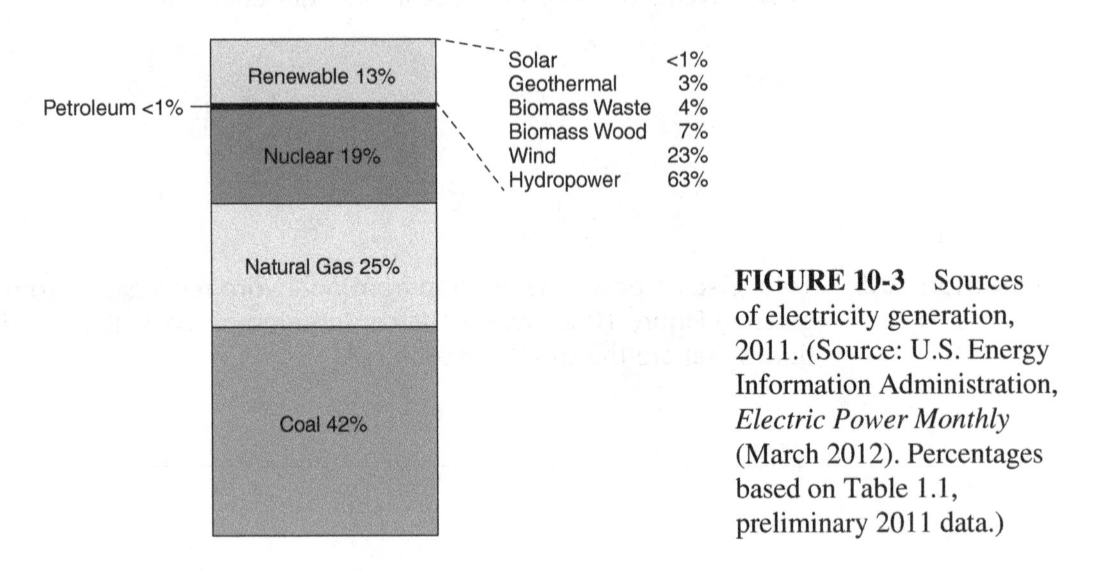

FIGURE 10-3 Sources of electricity generation, 2011. (Source: U.S. Energy Information Administration, *Electric Power Monthly* (March 2012). Percentages based on Table 1.1, preliminary 2011 data.)

Question 10-7: Total electricity consumption in the U.S. in 2011 was approximately 4190 billion kWh. What percentage of this was from wind?

Question 10-8: Discuss whether wind energy can make a significant contribution to U.S. electricity use in 2020.

Can Sustainable Sources Provide the U.S.'s Total Electricity Demand?

Recall that the U.S. consumed over 4 trillion kWh of electricity in 2011. To evaluate whether sustainable sources could meet the U.S. electricity needs, you will have to consider among other things the limitations of sustainable systems. We will focus here on solar and wind power. Keep in mind that we need electricity 24 hours a day and that there are few ways to store electricity once it has been produced. And recall that electricity demand varies over the day.

Question 10-9: Based on your typical day, draw a chart estimating when electricity demand is greatest and least. What are some limitations of solar and wind power?

For our analysis of sustainable energy, we focus on the present and future of wind power systems.

WIND ENERGY DEVELOPMENT IN THE U.S.

After a decade of more or less frenzied development, by early 2012 the U.S. wind industry had more than 48,600 megawatts (MW) of installed capacity, and had another 8,900 MW under construction, according to the American Wind Energy Association.[3]

Question 10-10: It cost about $3.5 million to install a typical 2 MW wind turbine in 2012. What was the average cost to install one kilowatt of wind energy?

Question 10-11: How does this compare to the cost of $700 per kilowatt to build a natural gas-fired plant?

Question 10-12: What advantage does wind have over gas that could offset this higher installation cost?

[3] American Wind Energy Association, http://www.awea.org/learnabout/.

Wind "farms" continue to be second-largest source of new power generation built in the United States, after new natural gas power plants. The United States as of 2012 ranked third in the world in wind energy, behind Germany and Spain.

According to the industry's trade association, a 1.5 MW wind turbine situated optimally should generate over 4 million kilowatt hours per year—enough to supply 400 average homes.

Question 10-13: Based on information in the paragraph above, what is the annual electricity consumption of an average house in the United States?

Question 10-14: According to General Electric, one of the largest turbine manufacturers in the world, 2.5 MW turbines are beginning to be installed globally. How many typical homes could be supplied by a 2.5 MW turbine, compared to a 1.5 MW turbine?

Question 10-15: According to the U.S. Energy Information Administration EIA, the United States consumed about 3,800 billion kilowatt hours of electricity in 2011. How does this compare to the wind energy potential of the top five states alone in Table 10-1?

TABLE 10-1 ■ States with the Greatest Wind Potential
The top twenty states for wind energy potential, as measured by annual energy potential in billion kWh.

Rank	State	Potential	Rank	State	Potential
1	North Dakota	1,210	11	Colorado	481
2	Texas	1,190	12	New Mexico	435
3	Kansas	1,070	13	Idaho	73
4	South Dakota	1,030	14	Michigan	65
5	Montana	1,020	15	New York	62
6	Nebraska	868	16	Illinois	61
7	Wyoming	747	17	California	59
8	Oklahoma	725	18	Wisconsin	58
9	Minnesota	657	19	Maine	56
10	Iowa	551	20	Missouri	52

(*Source:* www.energyonline.com/Restructuring/news_reports/news/1011ren.html. Courtesy of Energy Online.)

Question 10-16: Based on these data, could wind supply the country's entire electric power demand, assuming enough turbines could be built? Why or why not? What additional information would you need to more fully answer this question?

Some Aspects of the Economics of Wind Energy

The cost of wind power is strongly affected by three factors: average wind speed (wind power increases as the cube of its speed), interest rates (since companies typically borrow money to build the plants), and any government wind subsidy, which averaged 1.9 cents per kilowatt hour for the 2000s. The federal Wind Production Tax Credit (PTC) was first passed in 1994 and enhanced the economics of wind production through a tax incentive. It provided 1.9 cents per kilowatt hour generated for the first ten years of a wind turbine's output. The Act must be reauthorized every two years by Congress, leading to uncertainty in the future potential for wind power systems.

Even though there is no fuel cost for a wind plant, operators often pay royalties to landowners, and such plants are relatively expensive to build; thus, much of the cost is the capital required for equipment manufacturing and plant construction. This, in turn, along with the absence or presence of the PTC, means that the economics of wind is unusually sensitive to the prevailing interest rate, and politics.

Moreover, as with any form of electricity, the power must be shipped by transmission line from the generation site to the site of consumption. This limits the distance that electricity can be efficiently transported. It is impractical to produce electricity in California and ship it to New Hampshire for consumption, for example.

Question 10-17: The new method of hydrofracturing, or "fracking," has substantially increased natural gas reserves in the U.S. (see Chapter 9), and substantially lowered the price, from around \$6 per thousand cubic feet (MCF) to less than \$3. What effect do you think this price decline has on the economics of wind development?

FOCUS: RENEWABLE ENERGY IN HAWAII

Alone among the fifty states, Hawaii cannot call upon its neighbors for electricity. As a result, the state is actively supporting the development of renewable energy resources. Imported oil as of 2012 supplied more than three-quarters of Hawaii's energy needs.[4] No other U.S. state is so critically dependent on imported oil. Environmental and energy security concerns, however, make it essential to sharply reduce oil use in electricity generation and more effectively utilize local sources of renewable energy: solar, wind, water, wave, ocean thermal, biomass, and geothermal.

---

[4] www.hawaii.gov.

While as of 2004 the state obtained less than 0.1 percent of its electricity from wind and 2 percent from geothermal, by 2011 the state produced 12% of its electricity from renewables, and had a firm goal of moving to 40% from renewables (wind, solar, and geothermal) by 2030.[5]

Here is a state assessment of the attributes of renewable energy:[6]

1. Renewable energy projects create more jobs than a comparably sized fossil fuel plant.
2. A greater use of Hawaii's abundant renewable energy resources would help to insulate the state against fossil fuel price escalation and supply disruptions.
3. Money spent on indigenous energy would largely remain in the state.

There are also substantial environmental advantages to renewable energy. Reducing oil use would reduce oil spill risks. This is important to the state's tourist industry, as well as to its extensive marine nature reserves.

But there are also limitations. These include the high cost of land, as well as technical limitations in the state's electricity distribution system. Here is the state's assessment of the overall potential.

Renewable energy projects can theoretically provide all *new* generation required to satisfy projected electricity demand increases in the state through 2014. On Oahu, where most people live, using optimistic assumptions, renewable energy projects could meet all of the electricity required to meet projected demand.

The assumptions usually refer to reasonable interest costs, land lease costs, zoning requirements, construction costs, etc. Although the state considers that wind projects might displace agriculture and thus might not be viable, many wind sites are presently operated on sites compatible with agricultural activity. However, in Hawaii, the two main crops, sugar and pineapples, require harvesting techniques that are not as compatible with wind energy as other crops. With pineapple cultivation, for example, the location of wind turbine towers could interfere with the equipment used to harvest pineapples. The burning of cane fields could also damage wind turbines, but this also represents a biomass residue that could be burned for fuel or converted to ethanol.[7]

Some of the best sites for wind development are offshore. This could make it easier to obtain lease sites, assuming the tourist industry were to concur.

One factor that may ultimately decide the issue could be the future price of oil. Since the state report we cite here was issued, oil prices have nearly doubled, to around $100 a barrel, on world markets.

Question 10-18: What role should renewables play in a sustainable society?

[5] www.hei.com.
[6] http://hawaii.gov/dbedt/main/whats_new.
[7] Ibid.

Question 10-19: Summarize the important points of this chapter.

FOR FURTHER THOUGHT

Question 10-20: Another proven sustainable source of electricity is geothermal energy. Research and assess the state of geothermal energy development in the United States. Assess the future potential for expanding this resource.

Question 10-21: Solar photovoltaic technology employs another renewable energy source: the sun. Research the growth of the photovoltaic industry by searching the web, using the term "solar photovoltaics." Discuss the potential for photovoltaics development over the next several decades.

Question 10-22: When you drive or fly, you can offset your greenhouse gas emissions by purchasing greentags, also known as renewable energy certificates. The money you pay goes to develop alternative energy projects (typically wind or solar). The Green Power Network publishes a table listing greentag providers and comparing price premiums for residential greentags.[8] Some environmentalists consider greentags counterproductive, feel-good environmentalism. Research the issue and discuss why this is so and whether you agree.

Question 10-23: Hybrid cars are becoming popular in the United States, but the present generation permits few of the cars to be "plugged in" at night or when parked to recharge batteries. Experts claim this improvement could double city mileage for hybrids. Research this issue, and determine why manufacturers like Toyota are reluctant to sell cars that can be plugged in.

Question 10-24: Evaluate the extent to which wind potential is being used to generate electricity in your state or province. If your state or province does not have any wind developed, find out why. What sources produce your state's or province's electricity? Coal? Nuclear? Gas?

Question 10-25: Much future renewable energy development depends in part on the future world price of oil and natural gas. You have seen what has happened to natural gas prices in the U.S. due to "fracking" production. How predictable is the future oil price? Go to www.eia.doe.gov and determine how the government projects the world price of oil. Some have argued that oil-producing countries could sabotage the development of renewable energy by periodically dropping the price low enough to render renewables "uncompetitive" with oil. Is this a reasonable possibility? Why or why not?

---

[8] www.eere.energy.gov/greenpower/markets/certificates.shtml?page=1.

CHAPTER 11

THE THREE R'S: REDUCTION, REUSE, AND RECYCLING

KEY QUESTIONS

- How much municipal solid waste is recycled?
- What materials have the highest recycling rates?
- What is the difference between pre-consumer and post-consumer waste?
- Is recycling consistent with principles of sustainability?
- How do reuse and reduction differ from recycling?

RECYCLING

> "This is the most exciting thing that's come along for the last 15 or 20 years" in recycling, said Christine Knapp of Citizens for Pennsylvania's Future.[1]

Ms. Knapp was referring to RecycleBank, a nonprofit company started in 2004 by Patrick Fitzgerald and Ron Gonen. Here's how RecycleBank works: Households place recyclables in special bins that have a computer chip (essentially a barcode) embedded in them. Bins are weighed when they are emptied at curbside, and a scan of the barcode records which household the bin belongs to. RecycleBank Dollars, coupons that can be redeemed at various businesses including Starbucks, Home Depot, and Bed Bath & Beyond, or donated to charity, are awarded based on the weight of the recyclables. According to the company, there is virtually no cheating (by adding weight to the bins, for example). RecycleBank charges a fee to municipalities or private waste haulers, who typically offset this cost (or even make a profit) by reduced tipping (dumping) fees. They also receive revenue from some of the recyclable material. The program has tripled recycling rates in Philadelphia neighborhoods and had 3 million participants globally by 2012.

Question 11-1: Evaluate whether this is a useful approach to increasing recycling rates. List and evaluate other ways to increase the rate of curbside recycling.

[1] Philadelphia residents discover it pays to recycle. *Planet Ark,* www.planetark.org/dailynewsstory.cfm/newsid/35965/newsDate/11-Apr-2006/story.htm; www.recyclebank.com.

Recycling is also catching on at some special events:

- At the 2012 Phoenix Open Golf Tournament, sponsor Waste Management plans to remove trash receptacles and replace them with recycling and composting containers. It expects to divert 90% of potential waste from landfills.
- The Bonaroo Music Festival in Tennessee requires vendors to purchase compostable/recyclable paper products and cups.
- The U.S. Green Building Council's GreenBuild Expo works with a convention center, caterer, and hotels to implement programs to reduce, reuse, and recycle waste at meetings. Some stadiums and arenas are beginning to institute recycling at sporting events.
- In a 2010 survey, 65% of university athletic departments placed a "high" or "very high" emphasis on game-day recycling at athletic events.

Question 11-2: Does your institution or professional team recycle at games or special events? If so, do patrons use recycling bins, or do they merely throw away recyclables?

In 2005, Starbucks, in partnership with Environmental Defense, announced that new disposable paper cups would henceforth contain 10 percent post-consumer waste recycled content, a move that would result in savings of 5 million pounds of virgin tree fiber annually. According to Margaret Papadakis, head of packaging at Starbucks, the process took so long to implement because Starbucks' customers "expect their cups to look a certain way."[2] Starbucks used 4 billion paper cups worldwide in 2011, and proposes to ensure that all of its cups are "recyclable or reusable" by 2015.

Question 11-3: In 2004, Starbucks purchased 27,400 tons of 100 percent virgin-bleached cup stock.[3] By how much would landfill content have been reduced by the use of the new cups?

Table 11-1 shows the environmental benefits of using 1.9 billion cups with 10 percent recycled content, according to Environmental Defense.

TABLE 11-1 ■ Estimated resource savings associated with use of cups with 10% post-consumer waste content by Starbucks.[1]

- 11,300 fewer tons of wood consumed (or about 78,000 trees)
- 58 billion BTUs of energy saved (enough to supply 640 homes for a year)
- 47 million gallons of wastewater avoided (enough to fill 71 Olympic-sized swimming pools)
- 3 million pounds of solid waste prevented (equivalent to 109 fully-loaded garbage trucks)

[1]www.papercalculator.org

[2] www.sijournal.com/breakingnews/1919807.html.
[3] www.starbucks.com/csrnewsletter/winter06/csrEnvironment.asp.

To most Americans, being environmentally sensitive means one thing: recycling. In 2010, according to the EPA, recycling diverted 85 million tons of municipal solid waste from landfills and incinerators. Figure 11-1 shows the composition of MSW, and Figure 11-2 depicts the recycling rates for various materials.

Question 11-4: The 85.1 million tons of MSW that were recycled in 2010 represented a recycling rate of 34 percent. How much MSW was not recycled?

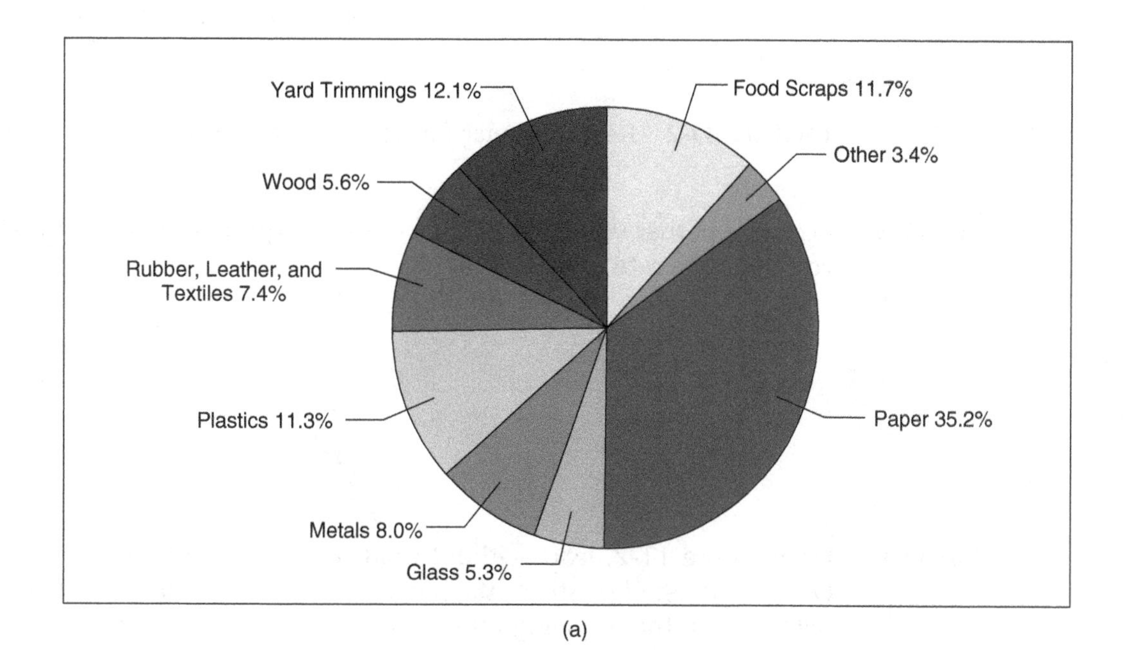

(a)

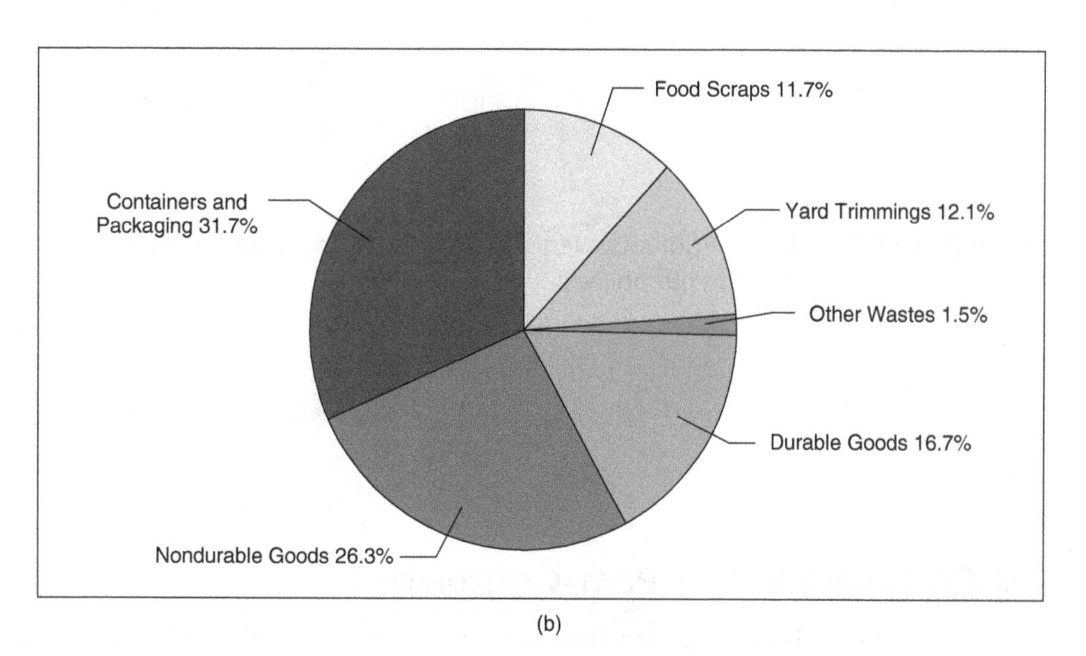

(b)

FIGURE 11-1 Composition of MSW in 2003. (a) Materials; (b) Products. (www.epa.gov)

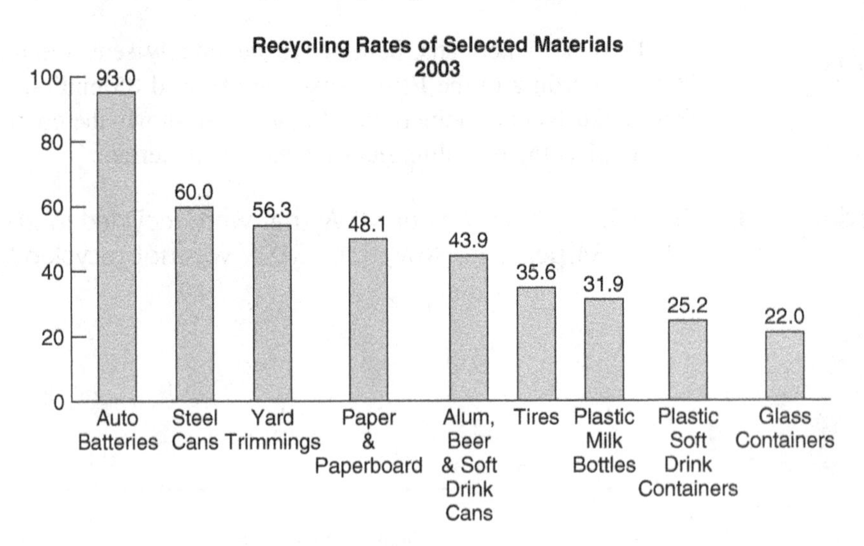

FIGURE 11-2 Recycling rates for selected materials in 2003. (www.epa.gov)

Question 11-5: Explain whether you think 34 percent is a high or low recycling rate. Discuss why you think the rate is not higher.

Question 11-6: From Figure 11-2, lead-acid auto batteries had the highest recycling rate, while two categories of plastic bottle had the lowest. What do you think accounts for the difference in the recycling rates?

Question 11-7: Would additional legislation help to increase recycling rates? Why or why not? Defend your answer.

PRE-CONSUMER VERSUS POST-CONSUMER

The average American consumes over 700 pounds of paper a year (the equivalent of nine telephone-pole size trees), nearly 90 percent of which is virgin (i.e., with no recycled content). Unless marked as containing 100 percent post-consumer content (see below), paper that is marked *recycled* also contains virgin fibers, but has some percentage of pre-consumer and/or post-consumer recycled content. Pre-consumer content refers to paper scraps and trimmings produced as a by-product of the paper manufacturing process. Post-consumer waste (PCW) is paper that has been previously used by the end-consumer. The higher the post-consumer waste content, the better the product is from an environmental standpoint. First, high-PCW paper diverts material from landfills, roadsides, or incinerators. In addition to saving trees, high-PCW paper uses significantly less energy and water to manufacture. Finally, many types of high-PCW paper are unbleached or use chlorine-free bleaching processes (which may also be used on some virgin paper products) that produce no dioxin. *Dioxins* are a class of persistent-carcinogenic compounds that are among the most toxic chemicals known (see Chapter 12).

Question 11-8: The University of Vermont created a paper purchasing policy in 1999 that recommended a minimum of 30 percent PCW content. Does your institution have a recycled paper policy? Check your institution's website, ask an administrative specialist, or call your purchasing department.

In the past, recycled paper did not meet minimum standards for use in high-speed copy machines and printers, but that has now changed.

Book publishers are starting to make the transition to high-PCW paper as well as other innovations. *An Inconvenient Truth,* by Al Gore, was printed on Appleton Green Power 80# Matte paper, which is chlorine-free (and is produced using 100% green power) and contained 30% PCW. The virgin component of Appleton Green Power paper was from sustainably grown trees. According to the *New York Times* (whose products contain on average 27% recycled content), Random House, whose "parent" Bertelsmann buys 3.8 million tonnes of paper annually, is moving toward using 30% recycled paper.[4] Random House's Chief Executive Peter Olsen said that the time has come "to stop making eye-rolling excuses about how burdensome and expensive environmental initiatives are."[5] Efforts by trade publishers and university presses to increase the use of high-PCW paper could decrease greenhouse emissions by over 250,000 tons annually and save 4.9 million trees, 2.1 billion gallons of water, and eliminate 1.3 million tons of solid waste.[6]

An alternative to both virgin and recycled paper is tree-free paper made from fibers such as hemp, sisal, abaca, kenaf, wheat straw, corn, and banana stalks. *Cradle to Cradle,* a book on sustainability by William McDonough, is printed on recycled (and, incredibly, recyclable) plastic.

Question 11-9: This textbook was printed on recycled paper. Is it clear what the PCW content is? Write down the titles and publishers of your other textbooks in the space below and what kind of paper (virgin, recycled, high PCW recycled) each was printed on. Ask your instructors if their decisions to adopt textbooks are based to any extent on whether the books contain recycled content, and report your findings here.

---

[4] Saving the planet, one book at a time, by Rachel Donadio, *New York Times Book Review*, 7-9-06, p. 27; www.bertelsmann.com

[5] Donadio op cit.

[6] Ibid.

E-WASTE

Question 11-10: What makes a computer or cell phone obsolete? Why do you "upgrade" your computer or cell phone? Do the added features enhance the quantity and/or quality of your work? Do they enhance the quality of your life?

In 2010, an estimated 14-20 million computers were thrown "away" in the United States alone. The National Safety Council reports that 250 million computers became "obsolete" by 2009. The "good" news is that legislatures, institutions, and individuals are starting to handle obsolete electronic equipment responsibly. Maine, for example, in 2006 passed a law requiring computer monitor and television manufacturers to pay for their recycling and disposal. Hewlett-Packard has two recycling plants that process 1.5 million pounds of electronics a month. And the manufacture of televisions with cathode-ray tubes (CRTs) is being drastically reduced as televisions with CRTs are replaced with flat-panel televisions. This saves about 8 pounds of lead for each CRT replaced with a flat-panel television.[7]

There is plenty of bad news as well. Elizabeth Royte, reports that many U.S. electronics recyclers simply ship the e-waste to foreign countries with less stringent or absent environmental regulations.[8] After usable materials are salvaged, the remains are dumped in landfills, along roadsides, or in water bodies. All of these dump sites can leach metals like chromium, tin, and barium at toxic concentrations into the ground and nearby streams. A report by the environmental organization Basel Action Network states that 50 to 80 percent of U.S. e-waste was being processed in China, India, Pakistan, and other developing countries under largely unregulated, unhealthy conditions.[9]

Question 11-11: If you knew that upgrading to a cell phone or computer meant that your old cell phone or computer would end up in a landfill or contribute to degrading the environment and harming the health of people in developing countries, would you still make the upgrade? Discuss whether you think this is a fair question.

[7] These and other initiatives are described in the *New York Times* article "Alternatives: Panning e-waste for gold," by Susan Moran (5/17/06).
[8] E-Waste At Large (NYT 1/27/06)
[9] http://www.ban.org/

Question 11-12: *New York Times* reporter Laurie J. Flynn quotes the president of Scrap Computers, a recycler in Phoenix, who stated, "There's no such thing as a third-world landfill. If you were to put an old computer on the street, it would be taken apart for the parts."[10] What do you think his point of view was? Re-examine then discuss whether you agree or disagree with his statement and his main point.

REDUCTION AND REUSE

An alternative to recycling is *source reduction*, which is the preferred method of waste management. Source reduction refers to buying less, buying more durable products, buying products with reduced packaging, and buying products designed to be reused.

Question 11-13: List consumer products that are designed to be useful for a short period (planned obsolescence), that are overpackaged, or that are not designed for reuse (after original use).

Question 11-14: If source reduction is the preferred method of waste management, why do you suppose the focus of most waste management programs is recycling?

THE FUTURE OF PLASTIC

Most plastic is a petrochemical product, some of which is recycled. One of the problems with plastic recycling (with the possible exception of William McDonough's book described above) is that it does not "close the loop." That is, unlike paper and aluminum, post-consumer plastic is not remanufactured into the same product that was recycled. For example, recycled soda bottles are not remade into soda bottles, but instead are manufactured into products that can use a lower grade of plastic (e.g., carpet, toys, or fleece). According to Eureka Recycling, one of the biggest non-profit recyclers in the United States, recycling plastic only delays its disposal. "The final destination for all plastic is either a landfill, where it doesn't decompose, or an incinerator, where it releases harmful chemicals when burned."[11]

---

[10] Poor nations are littered with old PC's. *New York Times*. 10/24/05.

[11] www.articleworld.org/index.php/Plastic_recycling.

Question 11-15: Major uses of plastic include disposable 2-liter bottles for sugary drink products and bottled water. Discuss the advantages of using plastic as containers for these products. Evaluate whether these are sustainable uses of plastic. List and discuss alternatives to these containers.

Question 11-16: Is shipping truckloads of bottled water thousands of miles using fossil fuels a sustainable practice? Could it be made sustainable? How?

Question 11-17: Considering that plastic production and recycling are fraught with environmental and environmental health issues, should we continue to use plastics as widely as we do? If not, discuss ways that plastic use could be reduced.

Another alternative to recycling is reuse, which includes personal choices like using travel mugs instead of disposable cups, and buying second-hand clothing and furniture.

Question 11-18: List additional examples of source reduction and reuse that you could adopt.

Question 11-19: Summarize the major points of this chapter.

FOR FURTHER THOUGHT

Focus: Aluminum Mining and Refining

Aluminum packaging is a mainstay of the beverage industry. Let's examine the environmental cost as well as the subsidies involved in the manufacture of the all-aluminum can.

Aluminum (Al) is one of the most common elements in the Earth's crust, but ores of aluminum are rare. The most common ore of aluminum is *bauxite*. Bauxite was first discovered in the French district of Les Baux in 1821, from which it took its name. Bauxite is found in extensive deposits in many countries, but mainly in the tropics and subtropics. Even though bauxite is a rich ore, the aluminum metal is tightly locked into the mineral structure and is not easily extracted.

First the bauxite must be crushed; then it is processed into a white powder called alumina (Al_2O_3). During processing, four tonnes of bauxite ore yield two tonnes of alumina and two tonnes of waste.[12] The two tonnes of alumina further reduce to one tonne of Al metal.

Question 11-20: How much waste is produced per tonne of Al refined?

The alumina is then dissolved in molten cryolite (Na_3AlF_6) to make the solution conduct electricity. An electric current separates the Al from the solution and molten aluminum metal sinks to the bottom of the vat, where it is drawn off. A by-product of the reaction, HF, is extremely toxic and is released as a gas. CO_2, a greenhouse gas, may also be emitted in the process.

Aluminum Cans

Cans are made from thin sheets of refined aluminum. The efficiency of can manufacture has been increasing. In 1972, 21.75 cans were made from a pound of aluminum. In 2008, that number had increased to 34.2.[13]

Aluminum melts at 660°C, so it's easy to recycle. It takes only 5 percent of the energy to make new cans from old cans (scrap) compared to what it takes to make new cans from bauxite, and of course, a great deal of waste is avoided as well (see above). For example, for each pound of aluminum produced by recycling scrap, 3 pounds of waste are not generated.

Consider these data:

Year	Recycling Rate (%)
1973	15.2
1974	17.5
1975	26.9
1976	24.9
1977	26.4
1978	27.4
1979	25.7

Question 11-21: Plot these data on the axes below, then project the recycling rate for 2010. (There are a variety of ways to do this: You can either extrapolate from your graph or calculate the most recent growth rate and use $N = N_0 \times e^{rt}$.)

[12] www.alcan.com.

[13] Aluminum Association. www.aluminum.org.

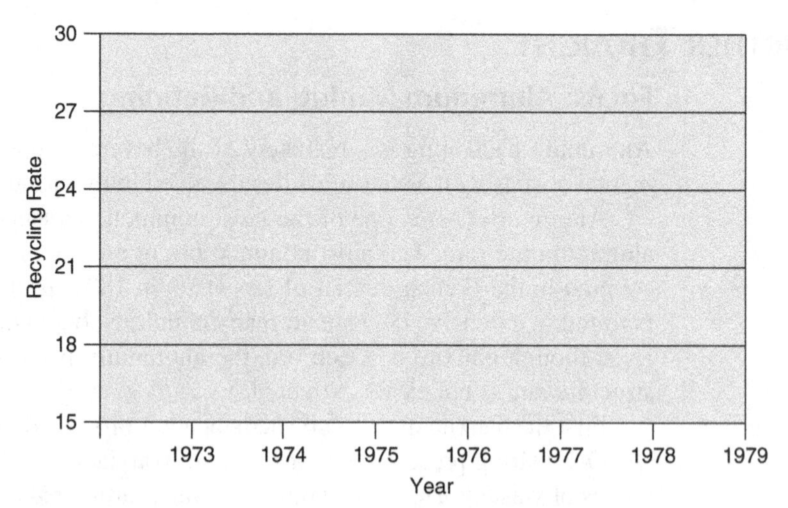

Here are the rest of the data:

Year	Recycling Rate (%)
1980	37.3
1981	53.2
1982	55.5
1983	52.9
1984	52.8
1985	51.0
1986	48.7
1987	50.6
1988	54.6
1989	60.8
1990	63.6
1991	62.4
1992	67.9
1993	63.1
1994	65.4
1995	62.2
1996	63.5
1997	66.5
1998	62.8
1999	62.5
2000	54.5
2001	55.4
2002	48.4
2003	50.0
2004	51.2
2005	52.0
2006	51.6
2007	53.8
2008	54.2
2009	57.4
2010	58.1

Question 11-22: Plot the recycling rates from 1972 to 2010 on the axes below. Did the recycling rate you projected from the 1972–79 data agree with the actual one? Did this reinforce or challenge your faith in the accuracy of projections? Why?

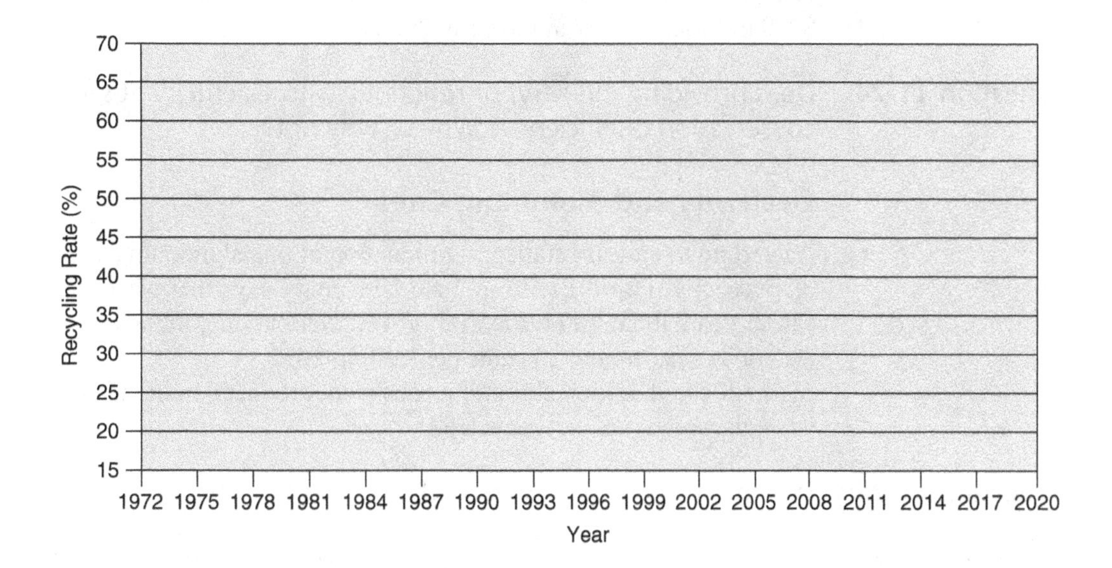

Question 11-23: On your graph, use a dashed line to project a recycling rate for 2020. How comfortable are you with your projection? Explain.

Subsidies and Aluminum Production

The electricity to extract the aluminum is only part of the cost to produce aluminum cans. Other costs include the extraction and shipping of the ore, the cost to restore the surface environment after mining is completed (or if restoration is not carried out, this cost is simply "dumped" onto the local environment and residents), the energy to ship the ore or concentrate to the refinery, and the cost to transport and distribute the cans. Economists define *externalities* as a cost that is not included in the price users are charged for a commodity. Subsidies are a type of externality.

Subsidies tend to distort markets, since subsidies encourage use of a commodity in excess of the level that the commodity would be used if all the costs to produce the commodity were included in the price. For example, the Australia Institute reports that the Australian government awarded mining companies subsidies of around $4 billion (Au) per year.[14]

Aluminum Refining and Electricity Costs

Refining aluminum from ore is extremely energy intensive, so most aluminum refiners either operate their own power dams, or concentrate in areas where electric rates are low. Such an area is the Pacific Northwest, where a federal agency established in 1937 by the Roosevelt administration, the Bonneville Power Authority (BPA), operates power dams on the Snake and Columbia Rivers. Under legislation that created the BPA, the agency must first serve public power agencies such as municipal utilities, as well as rural cooperatives and half a dozen government agencies. Some of what is left must go to investor-owned utilities. What is left over has been customarily sold to private industries, such

[14] The Australia Institute, https://www.tai.org.au/index.php?q=node%2F19&pubid=986&act=display.

as the aluminum companies. Where there was once plenty of leftover power, population growth in the Northwest and a booming high-tech economy has led to tight power supplies. By 2000, Bonneville was forced to go into the market and purchase about 1,000 megawatts of capacity. BPA provided subsidies to aluminum refiners (in the form of cheap power) at $200 million per year in the early 2000s.

Question 11-24: Does providing industry, or residents, with electricity at very low rates encourage conservation or efficiency? Why or why not?

Electricity and Aluminum Cans

According to industry statistics, production of one aluminum can requires about 0.35 kWh of electricity. David Biello of Yale University says that we make 300 billion aluminum cans a year, about 100 billion in the US. Nationwide, the average cost to *individuals* for electricity was around 10 cents per kWh in 2012.

In 2011, 98 billion aluminum cans were produced in the United States, and 42 percent were discarded (i.e., not recycled).

Question 11-25: How many cans were discarded?

Question 11-26: At 0.35 kWh per can how many kWh were thrown away with the aluminum?

The Sierra Club and the Competitive Enterprise Institute, among others, favor eliminating subsidies for electricity.[15] Many scientists urge increasing water flow around the BPA's dams to improve the rivers' environments for salmon and steelhead trout, which before dam construction supported a thriving fishing industry on these rivers. Some environmentalists favor dismantling many of the dams altogether.

Even though aluminum producers strongly support recycling and have underwritten thousands of recycling centers around the country (spending hundreds of millions of dollars in the process), the responsibility to deal with solid waste rests with local government, not with the producers of packaging. Can manufacturers point out that facilities to produce aluminum cans from scrap take half the time to build and cost one-tenth the amount of a mine and refinery.

Question 11-27: Is it fair to say that electricity used to produce thrown-away cans could be used for more productive activities? Discuss and explain your reasoning.

Question 11-28: Should producers of packaging be responsible for their product? Why or why not? Justify your conclusion.

Question 11-29: Do you think the destruction of fishing industries from rivers dammed to produce cheap hydroelectric power is a reasonable price to pay for artificially cheap aluminum cans that contain drinks with little or no nutritional value? Was this a biased question? Explain.

---

[15] www.cei.org

CHAPTER 12

PERSISTENT ORGANIC POLLUTANTS (POPS)

KEY QUESTIONS

- What are POPs?
- What is the "Dirty Dozen Plus Nine?"
- What international agreements address the problems posed by POPs?
- What quantities of POPs are in the environment presently, and how will this change in the future?
- What are PCBs, how were they formed, and what threats do they pose to the global environment?

THE CONVENTION ON PERSISTENT ORGANIC POLLUTANTS (POPS)

On May 23, 2001, U.S. officials signed the Convention on Persistent Organic Pollutants (POPs) in Stockholm, Sweden. Under the Convention, countries agree to reduce and eventually cease the production, use, and release of the twelve POPs of greatest hazard to the global environment. The agreement further sets up a scientific review process whereby additional chemicals may be added to the treaty as warranted.

The Stockholm Convention targets a group of POPs, informally called the "dirty dozen," shown in Table 12-1.

The POPs agreement begins global action to reduce and eventually eliminate these chemicals. The Convention took effect in May 2004 after ratification by fifty nations. By 2012, virtually every nation on Earth had ratified the Convention except Iraq, Saudi Arabia, Afghanistan, Italy, Malaysia, Western Sahara, Turkmenistan, Uzbekistan, and the United States of America.

In 2009, the Convention added nine more POPs to its list. They are:

Chemical	Source
Alpha-and-beta Hexachlorocyclohexane	Pesticide
Chlordecone	Pesticide
Hexabromobiphenyl and hexabromobiphenyl ether	Pesticide
Lindane	Industrial chemical
Pentachlorobenzene	Pesticide and industrial chemical
Perfluorooctane sulfonic acid and its salts	Industrial chemical
Endosulfan and isomers of same	Pesticide
Tetrabromodiphenyl ether	Industrial chemical
Pentabromodiphenyl ether	

TABLE 12-1 ■ The Dirty Dozen and Their Origins (Sources: EPA and World Bank).

1 = Pesticide 2 = Industrial Chemical	3 = Combustion Byproduct
Chemical	Comments
aldrin[1]	Fatal dose, 5g, adult male: fatal dose for women and children less.
hexachlorobenzene[1,2,3]	Can be lethal: has been found in food of all types.
chlordane[1]	Toxic to many animals. Human exposure is mainly by air.
mirex[1]	Used against fire ants. Very stable and persistent. Possible human carcinogen.
DDT[1]	Has been detected in breast milk. May harm infants.
toxaphene[1]	Most widely used pesticide in the United States in 1975. Possible human carcinogen.
dieldrin[1]	Mutagenic. Second most common pesticide detected in a U.S. survey of pasteurized milk.
polychlorinated biphenyls (PCBs)[2,3]	Suppress human immune system; probable human carcinogen; readily transferred in breast milk.
endrin[1]	Toxic, but can be metabolized, so little bioaccumulation.
polychlorinated dibenzo-p-dioxins (dioxins)[3]	Seven types out of 75 are mutagenic, carcinogenic.
heptachlor[1]	High doses fatal to birds, mammals; low doses mutagenic.
polychlorinated dibenzo-p-furan (furans)[3]	135 different types; possible human carcinogen; can accumulate in breast milk.

Question 12-1: Research one of these chemicals and report on its uses and toxicity.

WHY ARE POPS OF GLOBAL CONCERN?

The World Bank reports that, of all the pollutants released into the environment by human activity, POPs are among the most dangerous. POPs are of global concern because there is firm evidence of *global* transport of these substances, by air and water, to regions where *they have never been used or produced* such as the North American Arctic. The ensuing threats posed to the entire global environment prompted the POPs agreement. The main threats POPs pose center on their tendency to (1) persist in the environment, (2) bioaccumulate in the food chain, and (3) adversely affect human and animal populations. People are mainly exposed to POPs by eating contaminated foods. In humans and other mammals, POPs can be transferred through the placenta and breast milk to developing offspring. We discuss this phenomenon with *orca,* killer whales, below.

POPs are extremely toxic. They cause a range of harmful effects among humans and animals, including cancer, birth defects, damage to the nervous system, reproductive disorders, disruption of the immune system, and even death. POPs can damage the reproductive and immune systems of exposed individuals as well as their offspring. Some POPs are *endocrine disrupters.* An endocrine disruptor is a chemical that interferes with the function of the endocrine system. It can mimic a hormone, block the effects of a hormone, or stimulate hormone production. Table 12-2 shows persistence in agricultural soils of a number of POPs.

TABLE 12-2 ■ Persistence of POPs in Agricultural Soils (Courtesy of Ralph G. Nash and Edwin A. Woolson/AAAS.)

Chemical	Years since Treatment	% Remaining
Aldrin	14	40
Chlordane	14	40
Endrin	14	41
Heptachlor	14	16
Toxaphene	14	45
Dieldrin	15	31
DDT	17	39

COSTS OF EXPOSURE TO POPS

In June 2000, the National Academy of Sciences, the federal government's premier science advisory organization, issued a report titled "Scientific Frontiers in Developmental Toxicology."[1] Here is a portion of the Executive Summary:

> "Of approximately 4 million births per year in the U.S., *major* developmental defects [occur] in approximately 120,000 *live-born* (our emphasis) infants."

Question 12-2: According to the NAS, what percent of live births results in major developmental defects?

> "At present, the causes of the majority of developmental defects are not understood; however, it *is* known that prenatal exposure to certain chemicals (such as POPs) and physical agents (such as radiation) found in the environment can cause developmental defects. Approximately 3% of all developmental defects are . . . caused by exposure to toxic chemicals and physical agents, including environmental agents, and almost 25% of all developmental defects might be due to a combination of genetic and environmental factors."

It is no longer controversial that environmental toxins impose a significant cost upon humans, especially fetuses and infants. If we were to assume a minimal cost of each human life at $1 million (based on awards in liability lawsuits), then the *minimal costs* of such toxins in the environment could exceed $3.6 billion annually in the United States alone, and is likely to be much more."

POPS, ORGANOCHLORINES, AND THE PRECAUTIONARY PRINCIPLE

Recall the precautionary principle from the first section of this book. Restated, it reads, "Human society should avoid practices that have the potential to cause severe damage, even absent absolute scientific proof of harm." There is no better example of the need for application of the precautionary principle than that group of POPs produced when chlorine gas reacts with organic compounds. The compounds produced are called *organochlorines.*

[1] National Research Council: Committee on Developmental Toxicology, Board on Environmental Studies and Toxicology, Commission on Life Sciences. 2000. *Scientific Frontiers in Developmental Toxicology and Risk Assessment* (Washington, DC: National Academy Press.

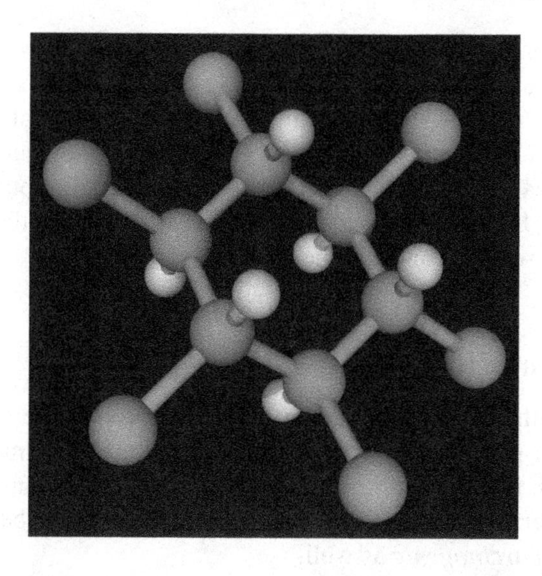

FIGURE 12-1 Molecular structure of an organochlorine. (Courtesy of Laguna Design/Science Photo Library/Corbis Images.)

Chlorine gas is greenish, heavy, extremely reactive, and does not occur in nature, not even in that cauldron of creation, active volcanoes. Chlorine gas readily reacts with any organic chemical it encounters, producing bleaches, disinfectants, insecticides and pesticides, and, incidentally, hundreds of additional by-products.

Organochlorines may be chemically organized in two fashions. One type, the *aromatics,* contains structures called benzene *rings* (see Figure 12-1). The other type, *aliphatics,* from the Greek for *fat* (fats have the chain structure as well), consists of *chains* of carbon atoms.

To give you an idea how these chemicals are named, consider PCBs. The chemical name for this family of compounds is *polychlorinated biphenyls.* A biphenyl is made of two linked benzene rings: If chlorines are added, it becomes a polychlorinated biphenyl. Some other extremely toxic chemicals are similar in structure to PCBs. DDT (*di*chloro-*di*phenyl*tri*chloroethane), for example, has one chlorine attached to each of two benzene rings, in turn, attached to a trichloroethane.

The aliphatics, since they closely mimic the structure of fats, are highly bioaccumulative (that is, they build up in fatty tissue). For example, hexachlorobutadiene has a bioaccumulation factor of up to 17,000. Others have ratios approaching 70,000.

Question 12-3: How many chlorines are found in hexachlorobutadiene?

How Does Chlorination Affect Organic Chemicals?

Chlorination radically changes the properties of organic compounds. It can increase the stability of the compound so that it may persist in the environment for decades or centuries. This property has been of great value to industry over the past hundred years and provided us with our first "safe" refrigerant, freon. The first generation of refrigerators used ammonia as a refrigerant, and ammonia is toxic when ingested. Freons were believed to be inert.

Since chlorinated hydrocarbons, by and large, are not naturally produced, there are few mechanisms that remove or degrade them once formed.

The addition of chlorine gas to organic compounds also increases their reactivity. And perhaps most important, adding chlorine to organic chemicals *increases their solubility in fats and oils.* This means they can bioaccumulate in fatty tissues of animals and can be passed on from generation to generation by mother's milk. They can also become concentrated in larger animals by the simple process of eating smaller ones, a process called *biomagnification.* Thus, humans can concentrate organochlorines in their bodies by eating contaminated fish, and seabirds and marine mammals can, and do, experience the same effect.

How Safe Are Organochlorines?

Only a few hundred of the thousands of organochlorines produced by industry have been tested by scientists, and virtually all of them have been found to damage one or more of the following processes: fetal development, brain function, the immune system, the endocrine system, and/or sperm production and development. They may be *mutagenic* (causes genetic mutations) and *carcinogenic* as well.

Organochlorines cause changes in bone composition including reduced bone mineral density. Organochlorines have also been implicated in other bone diseases including periodontitis, a widespread disorder of the gums and bones around the teeth.[2]

Question 12-4: Should women with a tendency for bone-density loss or people with periodontitis be encouraged to be tested for organochlorines? Why or why not? Do you think this is routinely done?

Finally all of these harmful effects can occur at *parts-per-trillion concentrations,* which is described as "equivalent to one drop in a train of railroad tank cars 10 miles long."[3]

Question 12-5: Reread Jefferson's quotation on corporations, from "Basic Concepts and Tools," (p. 15). Do you believe that corporations should be held liable for any damage that their actions cause, in violation of the precautionary principle? How? Cite evidence to support your view, recalling that opinions uninformed by evidence are of little value in scientific inquiry.

PCBs

Recall the "dirty dozen" from Table 12-1. One of these was polychlorinated biphenyls (PCBs). PCBs are a group comprising over 200 structures. Their formula is complex, and their atomic weight depends on the number of chlorine atoms in the structure. PCBs do not exist naturally on Earth. They were first synthesized during the late nineteenth century.

---

[2] National Institute of Health (NIH), www.pubmedcentral.nih.gov/articlerender.fcgi?artid=1331997.

[3] Thornton, J., in *Pandora's Poison: Chlorine, Health and a New Environmental Strategy.* 2000. (Columbia University, New York, New York.

Because of their stability when heated, they were widely used in electrical capacitors and transformers. In the 1960s, scientists began to report toxic effects on organisms exposed to PCBs, and by 1977, the manufacture of PCBs was banned in the United States, the United Kingdom, and elsewhere. By 1992, 1.2 million metric tonnes (2.6 billion lbs) of PCBs were believed to exist worldwide. As much as 370,000 metric tonnes (810 million lbs) could have been dispersed into the environment.

PCBs can be destroyed by incineration, but the process is expensive. Around 15 percent of PCBs in soils reside in developing countries, mostly from shipments of waste contaminated with PCBs from developed countries.

Question 12-6: What role, if any, should wealthy countries play in neutralizing PCBs in developing countries? Justify your answer by listing and defending your reasons.

PCBs are relatively heavy molecules (average atomic weight of around 360 g/mole) and are relatively insoluble in water. While concentrations in seawater can reach 1 part per million (ppm), PCBs typically concentrate in sediments. From there, they enter the food chain through the activities of organisms called sediment, or deposit, feeders. These creatures eat sediment, extract organic matter, and excrete the rest. Like other organochlorines, PCBs are *lipophilic* and thus tend to accumulate in the fatty tissues of animals. If other animals eat the deposit feeders, the PCBs are not metabolized and become more concentrated in the animal's fat (biomagnified). Concentrations exceeding 800 ppm have been measured in the tissues of marine mammals. According to the Environmental Research Foundation, this would qualify the creature for hazardous waste status!

PCBs are widespread pollutants and have contaminated most terrestrial and marine food chains. They are extremely resistant to breakdown and are known to be carcinogenic. PCBs have been linked to mass mortalities of striped dolphins in the Mediterranean, to declines in *orca* (killer whale) populations in Puget Sound, and to declines of seal populations in the Baltic.

While PCBs may threaten the entire ocean, the northwest Atlantic is believed to be the largest PCB reservoir in the world because of the amount of PCBs produced in countries that border the north Atlantic. PCBs have been shown to cause liver cancer and harmful genetic mutations in animals. PCBs may inhibit cell division, and they have been implicated in reduction of plant growth and even mortality of plants. According to a report edited by Paul Johnston and Isabel McCrea for Greenpeace UK,

> Since the rate at which organochlorines break down to harmless substances [has been] far outstripped by their rate of production, the load on the environment is growing each year. Organochlorines (including PCBs) are arguably the most damaging group of chemicals to which natural systems can be exposed.[4]

Bioaccumulation and Biomagnification of PCBs

POPs accumulate in the body fat of living organisms, as you have seen, and become more concentrated as they move from one creature to another. When contaminants found in small amounts at the bottom of the food chain biomagnify, they can pose a hazard to predators at the top of the food chain.

---

[4] Johnston P., & I. McCrea. 1992. *The Effects of Organochlorines on Aquatic Ecosystem.* London: Greenpeace International.

A 1997 study found that caribou in Canada's Northwest Territories had up to ten times the levels of PCBs as the plant matter on which they fed. PCB levels in the wolves that ate the caribou were nearly six times higher still.

J. Cummins, in a 1988 paper in *The Ecologist,* concluded that adding 15 percent of the remaining stock of PCBs to the ocean would result in the extinction of marine mammals.[5]

Highly Exposed Populations[6]

The long-distance transport of POPs toward the poles has contaminated the Arctic food web. Indigenous peoples of the Arctic experience a high intake of organochlorines from consuming a traditional diet featuring marine mammals, which have accumulated high levels of organochlorines from their food. Populations who have a diet rich in fish from contaminated waters, such as residents of the Great Lakes region (in the United States and Canada) and of the shores of the Baltic Sea, have a high intake of organochlorines. Children can have a higher intake of organochlorines than adults because of their comparatively high food intake. In addition, exposure of the developing young is of great concern because these stages of life are most vulnerable to the toxic effects of POPs. Nursing infants have a particularly high intake of organochlorines because of bioaccumulation in breast milk.

Question 12-7: Should pregnant women, or those desiring to become pregnant, be encouraged to be tested for organochlorine contamination? Why or why not?

Synergistic Effects: PCBs and Mercury

Combining exposure to toxic chemicals can multiply the harmful effects of each: This is called *synergism.* One particularly toxic mercury compound, methylmercury, biomagnifies powerfully as it goes up the food chain. Figure 12-2 shows this effect from the Florida Everglades. Water concentration of 0.1 part per trillion (ppt) was biomagnified to 2,000 ppt in plants, and so on. Mercury exhibits synergistic effects with PCBs and other POPs. A study of children born to mothers who consumed fish from Lake Ontario showed that prenatal PCB and mercury exposures interacted to reduce performance of three-year-old children on certain tests.

PCBs and Orcas in Puget Sound

Even though soldiers during World War II used them for target practice, orca have become a symbol of the Pacific Northwest. In 1999, Dr. P. S. Ross, a research scientist with British Columbia's Institute of Ocean Sciences, took blubber samples from forty-seven live killer whales and found PCB concentrations of 46–250 ppm, up to 500 times greater than those found in humans. Ross concluded, "The levels are high enough to represent a tangible risk to these animals."[7]

Ross compared the orca population he studied with the endangered beluga whale population of the St. Lawrence estuary of eastern North America, in which a high incidence of

---

[5] J. E. Cummins, The PCB Threat to Marine Mammals. *The Ecologist,* Nov–Dec 1988.

[6] www.greenpeace.org.

[7] "Toxin Threatens a Wonder of the Northwest," by M. L. Lyke Special to the Washington Post, Monday, November 8, 1999, Page A9.

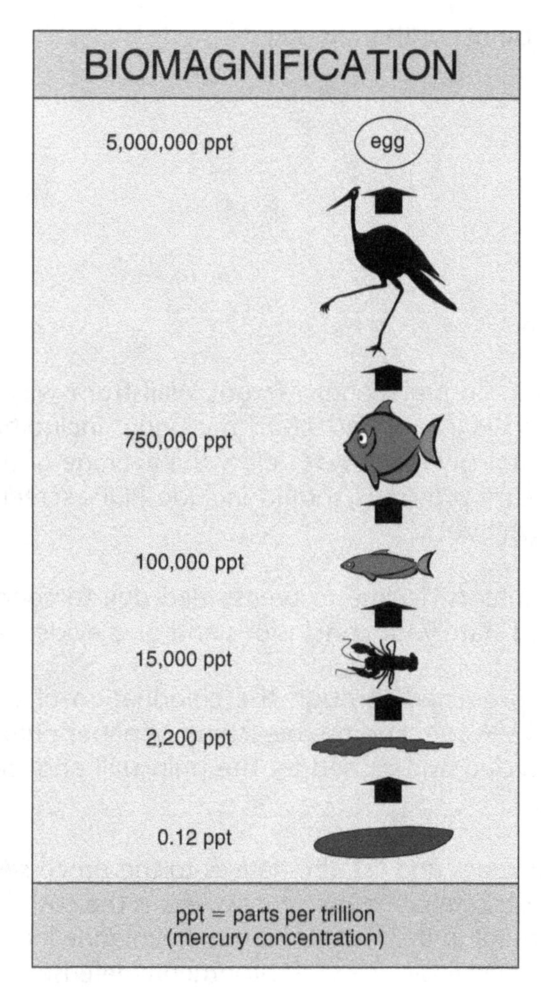

FIGURE 12-2 Biomagnification of methyl–mercury in the Florida Everglades. (www.usgs.gov)

diseases have been linked to contaminants and which have shown evidence of reproductive impairment.

For the orca, the PCBs are likely passed from generation to generation. PCBs are highly fat-soluble and are concentrated in mother's milk. Ross said, "Calves are bathed in PCB-laden milk at a time when their organ systems are developing and they are at their most sensitive."[8]

While PCBs have been banned in the United States for nearly three decades, they are still being used in some developing countries. Accordingly, Ross speculates that PCBs in the Pacific could be derived from East Asian sources and could end up concentrated in the tissues of migratory salmon, which are a prime food source for the Orca. Ross's study was done in collaboration with the University of British Columbia, the Vancouver, British Columbia, Aquarium, and the Pacific Biological Station of British Columbia.

Question 12-8: Discuss the topic of POPs from the perspective of sustainability and sustainable societies.

[8] "Killer Whales are Full of Toxic Chemicals, New Study Says PCBs Make Popular Orca Prey to Menacing Diseases." *Seattle Post-Intelligencer,* 10/25/99.

Question 12-9: Summarize the main points of this chapter.

For Further Thought

Question 12-10: European Environment Commissioner Margot Wallstrom was screened in 2003 for 77 toxins including POPs. "I had 28 in my body, including PCB and DDT," she said. "I was told that my result was below the average of the group tested."[9] Do you think health-care providers should include POPs screening for the entire population? Why or why not?

Question 12-11: Should women with infants decline to breast-feed due to concerns about POPs? Research this issue and state your conclusions and give evidence.

Question 12-12: Organochlorines are produced through the chlorination of drinking water, the purification of treated sewage, and the bleaching of paper pulp. Chlorinated pulp liquids cannot be recycled and reused by the pulp mill and must be discharged into local water bodies.

Research any of these issues and list alternatives to the processes used that rely on chlorine. Assess the costs to change the process versus the cost that is already being borne by the environment in the form of its organochlorine load. Consult the websites of Environmental Defense, www.environmentaldefense.org; the EPA, www.epa.gov; and those of major chemical producers like duPont for data on POPs. Also see *Pandora's Poison*, by J. Thornton (MIT Press, Cambridge, MA, 2000).

Question 12-13: Some critics assert that Greenpeace scientists (whom we have cited as references) may be too biased to provide reliable data on PCBs. Research PCBs on the Internet, in chemistry texts, or in other sources. Cite any differences between the treatment of these chemicals by Greenpeace and the sources you found.

Question 12-14: Here are some facts about the distribution of organochlorines:

Organochlorine compounds found in the tissues and fluids of the general North American population	193
Organochlorine contaminants so far discovered in the Great Lakes	83
Organochlorine by-products produced by the chlorination of drinking water and waste water	40
Organochlorine by-products of hazardous waste incinerators	31

Research these issues and conclude to what extent the hazards posed by organochlorines are offset by the value provided to society by the processes listed.

Question 12-15: Under the Convention (www.pops.int) may some POPs continue to be used? Why?

[9] www.cancerpage.com/news/article.asp?id=7077.

CHAPTER 13

SUSTAINABLE COMMUNITIES: *SPRAWL VERSUS SMART GROWTH*

KEY QUESTIONS

- How do scientists calculate population growth for a locality?
- What is sprawl, and how can we assess its environmental impacts?
- What are the implications of sprawl?
- How do municipalities deal with storm water?
- What is smart growth and what is it designed to accomplish?
- What are brownfields?

POPULATION GROWTH

> "Sprawl is not evil. In fact, it is good. It is the inevitable result of a free people exercising their cherished, constitutionally protected rights as individuals to pursue their dreams when choosing where to live, where to work, where to educate, and where to recreate."[1]

Population growth at the local level involves not only more people, but also high rates of land conversion, often called "development" The type of residential land conversion most common in the United States is *sprawl.* In this type of development, the amount of land converted from its traditional state (e.g., open space, farmland, forests) to residential and commercial use rises at a much faster rate than the growth of the population. For example, a study of urban growth in the greater Charleston, South Carolina metropolitan area from 1973 to 1994 found that over the twenty-one-year period, the rate of urban land use growth exceeded population growth by a 6:1 ratio.[2]

L. Brooks Patterson is certainly entitled to his opinion, but sprawl is associated with numerous kinds of environmental issues, not the least of which is transport emissions, especially CO_2. In a report done by the World Bank cities were ranked according to their population densities and CO_2 transport emissions per capita per year. Here is a brief snapshot of the findings.[3]

---

[1] The county executive of Oakland Co., MI, L. Brooks Patterson, http://www.oakgov.com/exec/brooks/sprawl.html.

[2] Allen, J., & Kang Shou Lu. Modeling and predicting future urban growth in the Charleston area. The Strom Thurmond Institute for Government and Public Affairs, Clemson University. www.strom.clemson.edu/teams/dctech/urban.html.

[3] *The Economist*, Shoots, greens and leaves; June 16, 2012, p. 69.

City	Population Density/ Hectare	CO_2 Transport Emissions Per Capita/Y*
Atlanta	2.4	7.5
Houston	14	6.2
Melbourne	16	2.5
London	50	1.2
Barcelona	160	1.6
Bangkok	185	0.8

Note: Barcelona and Atlanta have similar populations.
*Tonnes

Question 13-1: Describe the association between urban population density and transport emissions.

Question 13-2: Atlanta's metropolitan area is home to about 5.2 million people. How many tonnes of CO_2 per year could be saved if Atlanta was as efficient as Barcelona, with a similar population?

Prince William County, Virginia, located 35 kilometers south of Washington, D.C. along Highway I-95, has been one of the more growth-oriented entities in its region. From 1940 to 1990, Prince William's population grew at the rate of 4 percent per year. From 1990 to 2000, growth rates decreased to 2.6 percent. Its population in 2010 was 402,000 with a population density of 1,195 persons per square mile.[4] We will use Prince William County as our first case study of local growth. Between 2000 and 2007 alone, 9.4% of the entire county area was developed from open space/forest to commercial/residential structures.

GROWTH IN PRINCE WILLIAM COUNTY

When a population grows exponentially, the time it takes for the population to double, called doubling time, can be calculated using the formula $t = 70/r$, where t equals doubling time (usually in years) and r is the growth rate expressed as the percentage increase or decrease (for example, a rate of 7%, or 0.07, would be entered as 7). Doubling time is a very useful and practical concept for *projecting* (not *predicting*) and analyzing the implications of growth.

[4] http://quickfacts.census.gov/qfd/states/51/51153.html.

Question 13-3: Population growth from 2000 to 2010 was 43.2%.[5] What then was the average rate of growth over that ten-year period?

Question 13-4: The 2010 population of Prince William County was 402,000. Using the doubling time formula, determine approximately when the population of Prince William County would reach 1 million. Starting with the 2010 population, use the most recent percent annual rate (2000–2010).

Question 13-5: Now, let's apply the doubling time formula to project land-use changes. The area of Prince William County is approximately 9×10^8 m^2. At a 4.3 percent growth rate, when would Prince William County reach a population density of 1 person/100m^2?

Question 13-6: Answer the question again, using an earlier growth rate of 2.6 percent. Compare the two results and state how much more time it would take Prince William County to reach this high density with the lower growth rate.

Question 13-7: Discuss to what extent lowering the growth rate significantly reduces the impact of population growth.

[5] www.pwcgov.org.

IMPERVIOUS SURFACE

Development is an inclusive and somewhat ambiguous term used to refer to all the human-built structures in an area. One major effect of development is the increased amount of *impervious surface.* Impervious surface refers to any surface material that water cannot penetrate. In this section, we will analyze impervious surface as a local population growth issue.

When most people speak of development, they mean subdivisions, commercial buildings (such as offices, shops, and malls), and roads. Most of these buildings have parking areas attached. The parking area can be as small as the driveway of a house or it can be a paved area adjacent to a large mall, covering tens of thousands of square meters. These impervious surfaces collect runoff and prevent it from infiltrating into soils and surface sediment, where rainfall can be stored and natural filtering can often remove some pollutants.

Paved areas, and the vehicles that are parked on them, can contribute significant amounts of pollution to the water. But even if the runoff were to contain no pollution, it still would increase the risk of flash flooding. Local government officials are familiar with these threats and try to design stormwater management systems to handle runoff from development. Most municipal systems consist of a network of pipes that collect runoff from streets and large parking lots and channel it into artificial stormwater detention ponds or into creeks in the vicinity. Sometimes, the runoff pipes carry the water to a sewage treatment plant, where it can be treated before it is discharged into streams.

Stormwater management systems can, but rarely do, incorporate *Better Site Design* principles, including reshaped zoning regulations, increased green space, pervious concrete, vegetated swales and buffers, rain gardens, narrower streets, and even vegetated rooftops.

Although treating runoff to remove pollutants is a good idea, it can cause serious problems during times of heavy runoff. The added runoff from the stormwater system can, and often does, overload the sewage treatment plants, resulting in a mixture of untreated sewage and storm water dumped into waterways. For this reason, many communities design separate systems to transmit sewage and stormwater. Although this is more expensive, it results in much less environmental degradation, especially during floods. Unfortunately, it can also result in polluted water from impervious surfaces contaminating waterways.

Stormwater management systems work only as well as the weakest link in the chain, which simply means that if anything can go wrong, it usually will. Pipes and retention ponds may get clogged with debris or fill up with vegetation and sediment. Storm drains may be too small to handle runoff volumes, and development since construction of the system may overload them. This means that unless local governments take painstaking care in the design and maintenance of stormwater management systems, a costly endeavor that cash-strapped cities often forego, the more development in an area, the higher the risk of flash flooding.

HURRICANES AND TROPICAL STORMS

A hurricane or tropical storm can often dump as much rain on an area in a few hours as the area experiences in several months. Consider this example from Prince William County. The Potomac Mills Mall is one of the largest in the United States, and in 1999 the buildings and parking area together represented 5,675,411 ft^2 of impervious surface.[6]

When the last great hurricane, Agnes, hit the area in June of 1972 and dumped 14 inches of rain on the region in three days, Agnes' effects were described by officials as

[6] Prince William County Department of Public Works (Prince William County did not publish impervious surface data for 2000 and later.)

the Eastern Seaboard's most costly disaster in history. But Potomac Mills Mall hadn't yet been built.

Question 13-8: Assume another Hurricane Agnes-size storm occurs (as it surely will someday) and dumps 14 inches of rain on the region in 72 hours. Assume further that before Potomac Mills Mall was built the land was pasture and forest, which, given the area's subsurface geology, could have absorbed much of the storm's runoff. Calculate the maximum amount of extra runoff Potomac Mills Mall would generate when the next "Hurricane Agnes" hits northern Virginia. Express your answer in cubic feet and liters (1 $ft^3 = 2.83 \times 10^{-2}$ m^3; 1 m^3 = 1,000 L).

Question 13-9: Assume this runoff has to be handled over a 72-hour period. What is the resultant discharge from Potomac Mills Mall in liters per hour? (For comparison, a typical summer flow of the Potomac River is 100 million L/hr.)

Of course, this number represents a maximum value, but it illustrates how development can increase a region's susceptibility to flash flooding. Keep this point in mind while you study the impact of home building on Prince William County's impervious surface in the following section.

HOME BUILDING AND IMPERVIOUS SURFACE

Table 13-1 contains data on housing units and impervious surface for Prince William County from 1940–2010.[7] The area of the county is 222,615 acres or 90,000 hectares.

TABLE 13-1 ■ Prince William County's Housing Data Impervious Surface Area[1]

Year	Housing Units	Impervious Surface Area (m^2)	Area of county covered by impervious surface (from housing alone) %
1940	3,545	622,018	0.07
1950	5,755	1,009,850	0.11
1960	13,207	2,317,519	0.25
1970	29,885	5,244,050	0.58
1980	46,490	8,157,855	0.90
1990	74,759	13,107,552	1.45
1994	90,759	15,175,716	1.68
2000	94,570	—	—
2002	108,004	—	—
2005	123,379	—	—
2010	137,115	—	—

[1]Source: Prince William County Department of Public Works.

[7] Prince William County Department of Public Works.

Question 13-10: On the first set of axes below, plot housing units versus time. On the second set, plot impervious surface area (m^2) versus time. (Use data from 1940–2010.) Interpret the graphs.

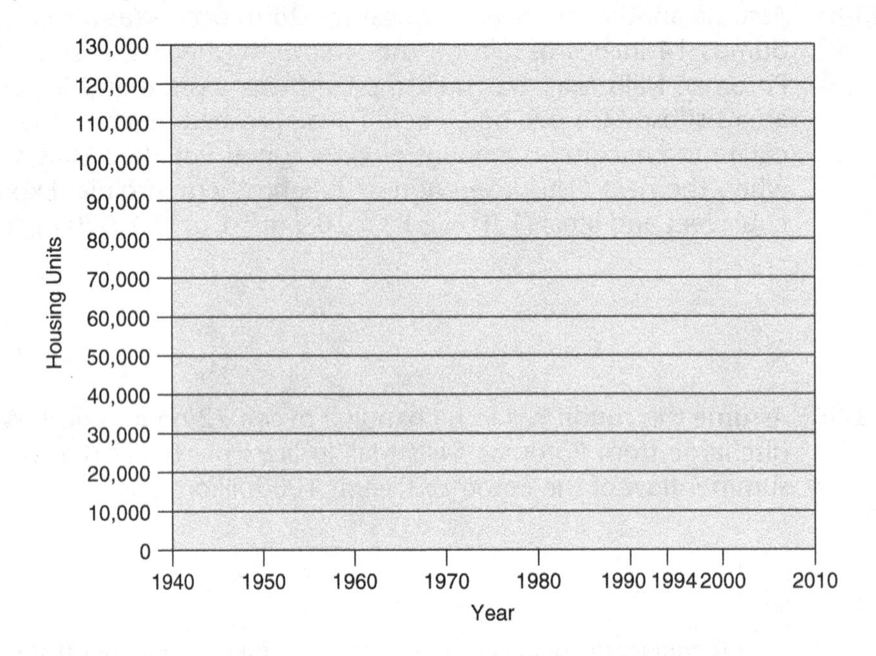

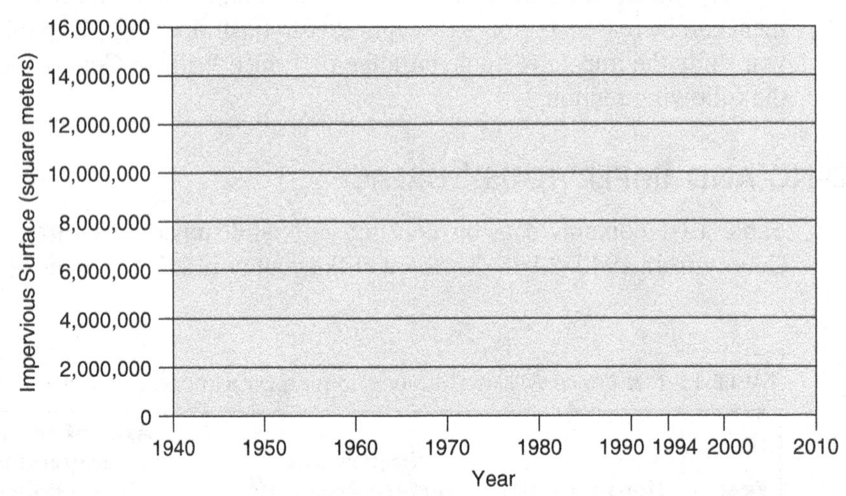

(To answer the following questions, use the equation, $r = (1/t)\ \ln(N/N_0)$, explained in "Using Math in Environmental Issues," pages 6–8.)

Question 13-11: Calculate average annual growth rates for both housing units and impervious surface area for the period 1940–94. How do they compare?

Question 13-12: Using the growth rate for impervious surface area from 1940 to 1994, project the area of impervious surface as of 2010. How did your calculation compare to your projection?

Question 13-13: Speculate how life would be different in Prince William County if half the county were impervious surface.

Question 13-14: In Question 13-4, we asked you to project when Prince William County's population would reach a density of one person per 100 square meters, which is certainly a very high density. It has been observed that population growth will stop eventually. We can decide when, or we can let math and "nature" take its course. Comment on why no one seems to be planning for extremely high population densities.

Question 13-15: Discuss whether "sustainable growth" is an oxymoron.

Question 13-16: Summarize the main points of this chapter.

FOR FURTHER THOUGHT

Question 13-17: The Chamber of Commerce generally opposes growth controls. Speculate as to why this might be so. Log on to a state or local Chamber of Commerce website, and then summarize and critically evaluate their position on growth controls.

Question 13-18: Log onto the Sierra Club's website (www.sierraclub.org/sprawl/). How do they explain their position on sprawl and local population growth? Critically assess it.

Question 13-19: Advocates of unrestricted growth frequently argue that growth expands the tax base, which lowers the tax burden for everyone and thus improves the quality of life. However, the results of more than seventy community studies conducted by the American Farmland Trust indicate for every dollar of revenue generated by residential development, local governments incur a median cost of $1.15 to provide services, and costs can be much higher. By comparison, the median cost of providing municipal services to farm, forest, and open land is $0.35 per dollar of revenue generated. The median cost of providing services to commercial/industrial development is $0.27 per dollar generated.[8] Research and discuss this issue.

Question 13-20: Find U.S. Census data (www.census.gov) or go to another source for data on growth for your county and make similar calculations to those we did for Prince William. (The California Department of Finance has excellent statistics, for example.)[9] What is the growth rate? What are its implications?

Question 13-21: Consider transportation. In 2012, there were 756 motor vehicles per 1,000 persons in the United States. Assuming the number of motor vehicles grows with the population, assess the impact of population growth on road congestion in your county or city. Is road construction and maintenance an unavoidable cost of population growth? Should people who choose not to own cars pay for these new or expanded roads? Provide evidence for your conclusions.

Question 13-22: Examine your local zoning ordinances, which you can usually find at your city or town planner's office. Are they designed to promote or limit growth?

Question 13-23: Study articles from your local newspaper or from a metropolitan newspaper in your library that focus on growth. Are growth rates given? If not, can you calculate them? Do the articles describe growth as "healthy" or "robust"? Do you agree that growth is "healthy" or "robust"? Discuss. Many of these articles refer to high population growth as "population booms" or "baby booms." What is the connotation of these descriptions? Can you think of a term that is neutral or more appropriate?

Question 13-24: To find out what your state is doing about growth, go to www.smartgrowth.org/. Discuss whether your state is taking actions to promote or discourage growth.

---

[8] www.farmland.org.

[9] California Department of Finance, www.dof.ca.gov.

INDEX

CREDITS

PHOTO CREDITS

6.CO Gerry Ellis/Digital Vision/Photolibrary; **6.2** Wolfgang Kaehler/CORBIS; **6.3** Victoria McCormick/Animals Animals - Earth Scenes; **6.4** Ken Hammond/USDA; **6.7** Edward Kinsman/Photo Researchers, Inc.; **6.9** Bryan Smith

FIGURE CREDITS

Fig 6.8 Images courtesy of Joseph Tomelleri, Virginia-Maryland Regional College of Veterinary Medicine.